Manufacturing Information & Data Systems

Acknowledgements

I am grateful to and wish to acknowledge Dr Jean-Noel Ezingeard for suggesting the initial structure for this book.

Manufacturing Engineering
Modular Series

Manufacturing Information & Data Systems

Analysis, Design & Practice

Franjo Cecelja

Penton
Press

Dedication
To Ankica, Lidija and Marina
Thank you

First published in 2002 by

Penton Press
an imprint of Kogan Page Ltd
120 Pentonville Road
London N1 9JN
www.kogan-page.co.uk

© Franjo Cecelja, 2002

British Library Cataloguing in Publication Data
A CIP record for this book is available from the British Library

ISBN 1 8571 8031 3

Typeset by Saxon Graphics Ltd, Derby
Printed and bound in Great Britain by Biddles Ltd, Guildford and King's Lynn
www.biddles.co.uk

Contents

Preface

Today's manufacturing is more than ever governed by consumer demand forcing companies to adopt a flexible and efficient way of operating. Information technology has proven to be a very useful tool in accomplishing this task – so much so that it has rapidly permeated manufacturing organizations at every level. There is a growing need for those related to manufacturing and associated business to understand the nature of this technology and the way it can best be harnessed to provide information for manufacturing functions and hence to increase competitive advantage. This book aims to explain:

- the nature of manufacturing information and its importance for a company;
- generation, manipulation and use of manufacturing information;
- technologies that help in generation, manipulation and use of manufacturing information;
- strategic implication of information systems on manufacturing enterprises.

It is impossible to appreciate the way that information and concomitant information systems can aid the competitive advantage unless a basic understanding is obtained of the manufacturing information itself, and also of the information technology and human aspect related to its use. The level of understanding needs to be comprehensive to enable those in manufacturing to assess the importance and also the opportunities and limitations of modern information systems.

Chapter 1 introduces the concept of manufacturing information. After covering essential definitions of information, information technology and information systems, an overview of characteristics of manufacturing information is given and its classification explained.

Chapter 2 introduces database systems in general, and their application in manufacturing in particular. It begins with the database concept and by defining various database systems. The core of this chapter explains the design of relational databases and the process of design optimization – database normalization. Structured query language is also presented. The latter part of this chapter explains the application of relational databases in manufacturing and associated management issues.

Chapter 3 explains manufacturing resources planning systems that are commonly used today. It begins by explaining the material requirements planning concept, followed by extension to planning and finance issues. Implementation and management issues associated with manufacturing resources planning systems are also presented.

Chapter 4 focuses on shop-floor data collection systems as major generators of manufacturing information. It explains various technologies used for shop-floor data collection, their advantages and disadvantages. Human aspects related to shop-floor data collection systems are also covered.

Chapter 5 introduces the concept of telecommunications. The importance of telecommunications systems in their use for competitive advantage is also presented.

Chapter 6 explains the central ideas and concept of electronic commerce with focus on the use of the Internet.

Chapter 7 supplements previous chapters by explaining strategic implications of the use of information systems. Practical applications and consequences are at the heart of this chapter.

Glossary

1NF	the first normal form
2NF	the second normal form
3NF	the third normal form
4NF	the fourth normal form
ASCII	American Standard Code for Information Interchange
ATM	asynchronous transfer mode
AT&T	American Telephone and Telegraph
BCIM	Brunel integrated manufacture cell
BCNF	Boyce-Codd normal form
BOM	the bill of materials
bps	bits per second
CAD	computer-aided design
CAM	computer-aided manufacture
CCD	charge coupled device
CCMI	critical characteristic of manufacturing information
CCTV	charge coupled television (camera)
CIM	computer integrated manufacture
CMM	co-ordinate measurement machine
CODASYL	conference on data systems languages
CPU	central processing unit
CSF	critical success factors
DBMS	database management system
DBTG	database task group
DDBMS	distributed database management system
DDL	data definition language
DML	data manipulation language
EDI	electronic data interchange
EMI	electromagnetic immunity

EMC	electromagnetic compatibility
ERP	enterprise requirements planning
Gbs	gigabits per second
HTML	hypertext mark-up language
IMS	information management system
IS	information systems
ISDN	integrated services digital network
IT	information technology
JIT	just-in-time
Kbs	kilobits per second
LAN	local area network
Mbs	megabits per second
MIDS	manufacturing information and data systems
MIS	management information system
MRP	material requirements planning
MRP II	manufacturing resources planning
NCC	National Computing Centre
OCR	optical character recognition
OSI	operating systems interconnect
PBX	private branch exchange
PC	personal computer
PLC	programmable logic controller
RAM	random access memory
RS232C	serial link standard
SFDC	shop-floor data collection
SME	small and medium enterprise
SNA	systems network architecture
SQL	structured query language
TCP/IP	transmission control protocol/Internet protocol
URL	uniform resource locator
VAN	value-added network
WAN	wide area network
WWW	the World Wide Web

1

Manufacturing Information and Data Systems: General Concept

1.1 Information needs of manufacturing

1.1.1 An external point of view

The role of a manufacturing organization can be seen as the generation of resources, generally financial, to add value to raw materials, then called *finished products*, for which there is a demand. In order to achieve this transformation process, the manufacturing organization uses a number of resources in the shape of equipment, people, information, energy and finance. If, over the lifetime of the company, the cost of the resources used is lower than the revenue brought in by the resources generated, the company is seen as being successful. Consequently, from a macroscopic point of view, a manufacturing organization can be represented as a system at the intersection of a materials flow and a resources flow (Figure 1.1). The resources flow contains the information that is one of the basic needs of a manufacturing organization. Typical examples of information contained in this flow would be customer orders, legislation and request for payment. This information will in turn trigger further creation of information flows within the organization.

The information needs of a manufacturing organization are therefore clear from a macroscopic point of view since information constitutes one of the key external resources on which the organization has to rely in order to perform its functions. It can be

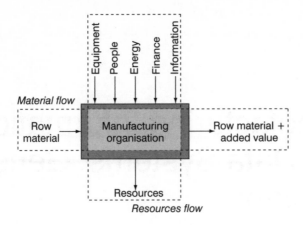

Figure 1.1 *Macroscopic view of a manufacturing organization*

assumed that in most cases the outgoing resources flow will also contain a certain amount of information, but other than in specific cases, this is not generally seen as revenue generating.

1.1.2 Information as a vehicle of integration

Since the creation of computer integrated manufacturing (CIM), information has been seen as an increasingly important component of manufacturing engineering, and as the most important vehicle of integration. Some authors describe manufacturing as having been successfully integrated with kinematics with the invention of the steam engine, with electricity on the invention of the dynamo, and now information with the arrival of computers (Broomhead, Grieve and Schmid, 1989). Information technology both speeds up information transfer and availability, and can therefore be seen as the main integrating factor in today's manufacturing environments. Manufacturing relies heavily on information as an important resource at all levels: from design through to planning, operations and after-sales operations. Information flows are the vital links between various manufacturing system elements.

The level of integration between manufacturing functions still varies greatly from one company to another, but, whether integrated or not, information is the lifeblood of an organization. In order to work as a team for achieving organizational goals, people want and need good information, both directional and feedback.

A practical example that shows the importance of information in manufacturing and its role as an integrating factor is the Brunel computer integrated manufacture cell (BCIM) shown in Figure 1.2. BCIM is the result of a teaching exercise of a final year option in CIM. It is a simple model of a manufacturing organization that uses industry standard equipment to bring together, on a small scale, all the functions of a typical production enterprise. It is based around a flexible manufacturing cell consisting of two machining centres, a co-ordinate measurement machine (CMM) and various trans-portation devices. The manufacturing cell communicates with the rest of the BCIM organization through a cell controller, which gets production orders from a production planning and control system called ACiT. ACiT uses data from a manufacturing resources planning (MRP II) application called *fourth shift*. The product defi-nition data are generated by the CAD/CAM (computer-aided design/computer-aided manufacturing) applications, currently AutoCAD and MasterCAM, which generate process plans from the CAD information. Product specifications arrive from the customer. The last functional area is that of computer-aided quality based around the processing of data from a CMM located in the cell. The data provided by the measurements of products can be analysed and the results used to offset the machine tools if necessary.

Figure 1.2 also shows the basic information flows in a BCIM based on a functional area representation of the organization. Different types of information clearly appear on this simple diagram. Some of the information is, for example, needed for logistical reasons such as schedules, whereas other information is needed for manufac-turing operations definitions such as process plans.

Manufacturing information will be described in greater detail in the next module, but this simple example is sufficient to show the dependence on information in manufacturing operations. No single functional area described could function without the infor-mation inputs described, which are broadly of two kinds:

- product information describing the material flow;
- control information defining the characteristics of this material flow and the operations taking place on the material flow.

The information needs of manufacturing, as highlighted above, are satisfied by subsystems of the manufacturing organization, which will be referred to as the manufacturing information and data systems (MIDS) throughout this book. Their functions range from order processing through computer-aided design and manufac-turing to production control. Computers can now be seen at work in

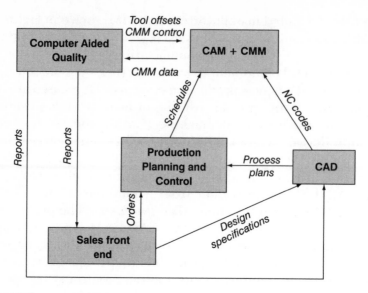

Figure 1.2 *Basic information flows between the functional areas of BCIM*

most functions of MIDS, such as order processing, design, manufacturing, logistics, accounting, etc. The level of integration between these functions always will vary greatly from one company to another because various different businesses and manufacturing processes will require different implementations. There is, however, a drive towards integrating all the functions that can be carried out in order to improve the efficiency of processes and their responsiveness. The consensus amongst manufacturing engineers is now that computer integration of production systems is relatively easy in technical terms. The management of the supporting information technology, from its selection to its introduction, is seen as the main area of concern. Issues of costs and benefits also need to be addressed, as well as selection of suitable technological solutions.

Manufacturing information systems are unusual insofar as, depending on the size of the organization using them and on the application required, they can be bought off the shelf, selected and implemented by consultants or wholly developed in-house. The choice is generally made by looking at a number of financial, operational, organizational and skill constraints. To some extent, the resources available for their development and implementation will also depend on the size of the organization, and this adds a further dimension to the management problems associated with MIDS.

1.1.3 History of manufacturing information systems

1.1.3.1 Financial evolution of production systems

Most authors in the field of manufacturing management talk about the new manufacturing environment of the present day, characterized by its competitiveness, its need for variety and therefore short product life cycles and cost structures shifting away from a high labour cost content. This new manufacturing environment evolved after the first oil crisis of 1973 when western economies moved away from overloaded production systems that could not satisfy demand due to lack of capacity.

After 1979, the markets started to be governed by consumers rather than producers and manufacturers became subject to increased competition on price, delivery, quality and technological evolution. This meant that more and more management information was needed about what was happening on shop floors in order for products to be produced to specification, on time and at the right price.

1.1.3.2 Technical evolution of computer systems

The financial evolution of the constraints on manufacturing was followed by a parallel technological evolution of computer systems. These were, until the 1970s, characterized by their very high cost, both in terms of hardware and software, which had to be developed in-house. As a result, only large companies were able to afford a computing department and the larger small and medium enterprises (SMEs) subcontracted part of their data processing.

The first computers to arrive in the manufacturing industry were confined to large companies who could afford the hardware as well as software development. Most of the companies set up a data processing department in charge of writing code and maintaining the equipment. The activities that were automated were those involving a large number of arithmetic calculations such as payroll, invoicing or accounting. The areas of application of computer systems before the 1970s were remote from manufacturing, which was not seen as a data-intensive activity. During the 1970s, with the appearance of mini-computers and the drop in price of hardware and improvement in performance of computers, this spread into industrial applications and robots became part of the industrial landscape.

The cost of computer equipment continued to fall in the 1980s, closely followed by the appearance of software packages on the market. This resulted in a large number of manufacturing applications of computers being available, with all manufacturing companies, including the smaller ones, becoming potential users of computers in manufacturing. Processing power became accessible to more and more companies, including SMEs that did not have the resources to develop and maintain their own software, but which were now able to buy software off the shelf. Company-specific software tended to make way for standard, cheaper products.

The 1990s to the present date is often referred to by the popular press as the *information age* and is, as far as manufacturing engineering is concerned, representing a move towards manufacturing communication at a global level and the total integration of manufacturing information systems and manufacturing data systems. The main technical factor behind this phenomenon is due to a combination in advances in telecommunications and computer networks.

1.1.4 Specific problems in today's manufacturing environments

The potential benefits of a suitable MIDS are great, but so are the potential costs. One of the main difficulties, as with any other technology, is for companies to choose what is appropriate for their needs and resources. Although much research work is currently being carried out to investigate the best ways of designing a manufacturing system, with simulation being seen as an important aid to machine selection and facility layout, the design of manufacturing information systems and the selection of equipment and software is still a difficult process. One of the main questions that needs to be answered is whether the system is going to provide value. Manufacturing companies are often unsure when deciding how far they should go down the route of integration of their current system. Table 1.1 lists a number of potential benefits and costs traditionally associated with MIDS.

Many organizations still see information technology only as an expense and manufacturing environments are lagging behind in deriving the full benefits that information systems have to offer (Gupta and Biegel, 1991). In other words, the left hand column of Table 1.1 is all too often ignored, resulting in information

Table 1.1 *A few potential benefits and costs of MIDS*

Potential benefits of MIDS	Costs/potential costs of MIDS
Competitive technology lead	Cost of equipment
Product quality enhancement	
Enhanced delivery performance	Loss of one-to-one contact with the customer
Production system flexibility	Cost of training
Efficiency and effectiveness gains	Loss of knowledge and know-how about the product
Readily available management information	
Social gains (quality of the work place, remote login, etc.)	Social costs

technology-related decisions based only on cost and necessity. Some authors point out that it has been estimated that more than 50 per cent of the costs of a manufacturing facility can be associated with an information overhead (Mitchell, 1991).

1.2 Information technology and information systems

Informatics is defined in the Oxford Concise Dictionary as the *science of processing data for storage and retrieval and also as all the techniques pertaining to the collection, sorting, transmission, and utilization of information*. These definitions are perhaps the best place to start in order to describe information technology and information systems, since they focus on information rather than the socio-technical systems that go around it.

An information system has, on top of the trivial information processing and usability function, four other important functions:

- education and learning function;
- information systems development function;
- management and control function;
- strategy and planning function (Jayaratna, 1994).

Although information technology and information systems are generally used interchangeably, the concept of information systems is broader than that of information technology. Information systems encompass the whole range of procedures that are in place in an organization and that deal with information, not only the automated functions (Figure 1.3). Improving the information system does not always mean improving the information technology. The reorganization of procedures and the redefinition of data paths yield immediate benefits and open up future opportunities for any company.

In a manufacturing environment, the technology needed to process the very large amount of information necessary for the production of parts is by nature varied. Information technology is generally described by its main hardware and software characteristics.

The final aspect of information systems is to take people into account. In any system that involves humans, information exchanges between humans are always the preferred form of communication. Further, it is a well-known fact that human networks are generally much more tolerant of flaws than computer networks because of their informal nature and capacity to integrate change informally. This simple fact is too often ignored when looking at information systems where the exact nature of information if created, transmitted or processed by humans simply cannot be determined. Therefore, information systems are considered to be a set of hardware, software, people and information exchange procedures dealing with all the operations on the information present in the system.

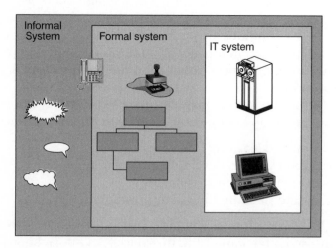

Figure 1.3 *IT as a part of IS*

1.3 Manufacturing information

Just as information is a difficult concept to define, what follows will show that manufacturing information, as one of its subsets, is also a complex notion particularly in the context of MIDS, which has been shown to be very varied. The times when Shannon could say that the semantic aspects of communication are irrelevant to the engineering aspects are long gone. It is crucial to understand information when analysing, designing, establishing, and managing information systems. Unfortunately, where people have concentrated on computer-based systems or allowed the requirements of data processing to become central to organizational structure, they usually lose sight of what information actually does for an organization.

A large number of definitions of information are available, which range from mathematical definitions to linguistic definitions. Some of these definitions include:

- the way a message content is structured;
- something that gives knowledge in the form of facts;
- element or system that can be transmitted by a signal or a combination of signals;
- measure of the density of intelligence contained in a message for a given number of signs;
- knowledge acquired through experiential study.

Nonetheless, understanding of information is seen as crucial by a very wide range of academics, from communications engineers to social scientists, and all these definitions serve the purpose for which they have been provided.

It is interesting to note that little thought seems to have been given to information specific to manufacturing. One argument for this would be that there are no grounds to consider manufacturing information apart from the broader concept of information. The following will, however, show that there are substantial characteristics of manufacturing information that can help the understanding of MIDS and therefore their management, and that this separation is helpful.

1.3.1 Definitions of information

Any management of technology activity has to rely on some tangible measures of success. In the case of information, there are a number of quantities that can actually be measured, some of which can be derived from mathematical definitions of information. Although some of those described below are of little practical applicability for this course, they are interesting insofar as they allow quantification, and therefore the potential measuring of value of the concept of information. They are, of course, also interesting from a historical point of view. Further, they lead to the first axiom of information physics, which is of some significance for this course.

1.3.1.1 Mathematical theory

According to this theory, the amount of information available is defined, as far as communication engineers are concerned, as the logarithm of the number of choices available when selecting a message (Shannon, 1949). The information content of a message can then be measured as its entropy, defined by:

$$H = -\sum_i p_i.\log_2 p_i \qquad (1.1)$$

where p_i is the probability of choice of symbol i. In the case of N equiprobable symbols for instance, $p_i = 1/N$ and therefore $H = -\log_2(1/N)$, as shown in Figure 1.4.

As mentioned earlier, this definition is relevant from a cultural point of view since it was the first time that information was considered to be a special entity. The definition is still useful for communication theory, but it must not be confused with meaning. Manufacturing information cannot easily be related to this definition since it only addresses the problem of communication and in fact only gives a measure of how well a message is organized. It relies

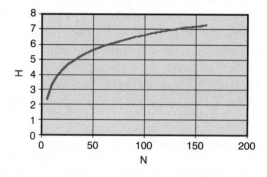

Figure 1.4 *Entropy of a source of N equiprobable symbols $H = -\log_2(1/N)$*

on the assumption that a symbol that does not appear often in a message gives more information than one that is more frequent. If the freedom of choice is reduced, the information is also reduced. This also means that information is the reduction of uncertainty, and further leads to the definition of a *bit* as the amount of information necessary to be able to chose between two alternatives.

1.3.1.2 Information physics

This definition is a more or less unique attempt to devise a theory of information. The theory is based on the following first axiom of information physics: *information and organization are intimately related* (Stonier, 1990). In turn, three theorems are then derived:

1. All organized structures contain information, hence no organized structure can exist without containing some form of information.
2. The addition of information to a system manifests itself by causing a system to become more organized, or reorganized.
3. An organized system has the capacity to release or convey information.

These theorems allow us to connect information by the well-known mathematical expression: $(-[entropy]) = k.\log(1/disorder)$, hence giving:

$$I = I_0 . e^{-\frac{S}{k}} \tag{1.2}$$

where I is a measure of information, I_0 is the information at zero entropy, k is Boltzmann's constant and S is the entropy. The relationship between the entropy S and information I is shown in Figure 1.5.

This equation obviously contradicts the mathematical definition (1.1): according to (1.1), a well-organized message conforms to rules that restrict the number of characters that can be chosen to make it up. The entropy (i.e., information content) is therefore reduced towards zero (see Figure 1.4). According to (1.2), however, the message would contain much information since it is well ordered and its entropy would therefore be very negative.

Manufacturing systems seem to be excellent examples of physical theory (1.1), where smooth running of the system and therefore good organization of the production process very much depend on the information exchanged between the control system and the production system. The order of a manufacturing system is, of course, linked to the information used for its control. The most important concept, at least for this course, is the axiom of information physics linking information and organization, and that information is not a construct of the human mind but a basic property of the universe. At this point, it could be argued that it is the key to understanding manufacturing information and data systems that deal with basic properties of the manufacturing systems, of which information is also a basic property.

1.3.1.3 Communication approach

The physical definition of information (1.1) introduces the concept of information machines that are able to convert energy into information (Stonier, 1990). The example quoted here is that of a radio transmitter, which transforms electricity into information. It is the case that the information output of a radio transmitter is the same as its information input, but it is multiplied by the number of receivers (Figure 1.6). This is where the physical theory (1.2) differentiates between physical information and human information. The human information is the same before and after transmission, whereas the number of receivers multiplies the physical information.

This concept of human information versus physical information is also useful to describe the phenomenon of information duplication by communication, which is an observation of the fact that, unlike traditional materials or resources, the movement of information does not cause it to disappear from one point or to be moved to another. It is, in fact, duplicated. If a customer has a need

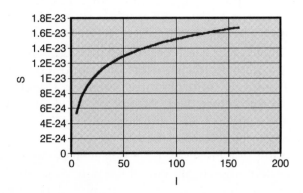

Figure 1.5 *Stonier entropy function*

for a product, which could be viewed as an item of information, and therefore sends an order to a supplier, seen as communication of the initial information, then this information now exists with both the customer and supplier. The supplier will then, in turn, change this information into a breakdown of production requirements, which will be passed on further in the supply chain.

It could be argued that at no point has the human information, as in the need for a product in previous example, been changed. It has, however, resulted in a multiplication of physical information, which cost resources to produce, unnecessarily perhaps, since the initial information is still present. This is, however, only true if, irrespectively of the number of duplications, the information remains the same throughout the process, which is, in fact, rarely the case. Communication causes human information to be transformed into data, which are then reinterpreted into information; but the information is not necessarily the same at both ends of the process.

Another interesting analysis is that which separates structural and kinetic information. In a manufacturing system, for instance, *structural information* is that which was provided at the system design stage by the engineers who specified and built the system and by those who designed its machines and procedures. *Kinetic information* is that which is produced, used and transformed during the operational stages of the system. This differentiation between kinetic and structural information is very relevant to MIDS since it will obviously influence the nature of the resources needed to deal with the information.

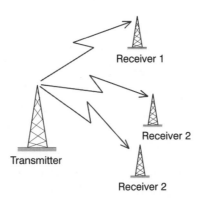

Figure 1.6 *Concept of communication*

1.3.2 Information science and infometrics

The American Society for Information Science was created in 1968 and groups together a number of scientists from a large number of disciplines. These include a number of applied sciences such as computer science and communication electronics, but also a number of social sciences such as communication and media studies. As with any other scientific area, there cannot be an information science without measures for information, and the American Society for Information Science has adopted five units that can be used in infometrics, as shown in Table 1.2.

These metrics are particularly useful in branches of information science such as communications theory, or in social sciences and bibliography where measures such as utility can be useful. The potential for their application in the study of MIDS is, however, very limited, perhaps being useful for low-level applications.

1.3.3 Systems approach to describing information

1.3.3.1 Concept

So far, information has rarely been defined in an operational sense. This idea is also at the base of the theory of information economics where data are perceived as a resource and managed as such.

An operational view of information is based on the fact that, in practice, information shows itself on all counts to be analogous to matter or energy, and an information network can be regarded to be similar to a material or energy flow, with similar components

Table 1.2 *Base units of infometrics*

Quantity	Symbol	Unit	Definition of example
Signal	S	bit	log_2 of the total number of possible states
Text	T	word	Smallest independent meaningful linguistic entity
Time	T	second	Duration of 9,192,613,170 periods of vibration of cesium 133.
Population	P	individual	Design engineer, production operative, etc.
Utility	U	usage	Number of uses and/or citations of an informational object

(Liebenau, 1990). This analogy can be taken further with the representation given in Table 1.3.

This representation does not describe various information elements; rather, it concentrates on information processing. Although limited, the representation helps in showing that defining information from an operational point of view is possible. There is, at least in principle, no major difference between an information system that would be represented using a source, sink, effector, reservoir and tap model and an electrical circuit diagram.

Information sources and information sinks are perhaps the most difficult elements to list exhaustively. At the information technology level, listing all inputs of information in a system is possible since the boundaries of such a system can be clearly defined. The boundaries of the human information system are, however, more difficult to define clearly from an input point of view. It would be impossible, for instance, to describe where the knowledge that a designer uses to draw a new concept comes from. In most organizations, the outputs of the human information system are easier to describe. Designers produce concepts, often in the form of drawings or sketches; process planners produce process plans, etc.

The conclusion can therefore be reached that an operational view, which can also be called a *systems approach*, of information and information processing is possible, but that in order to be rigorous, system boundaries need to be defined around the human information system, leaving some of its inputs or control mechanisms outside.

Table 1.3 *Analogy between energy, material and information flows*

	Energy (e.g., electrical)	Matter (e.g,. manufacturing)	Information
Source	Generator	Supplier	People, shop-floor data collection system
Sink	Environment	Customer	People
Effectors	Motors, resistors	Machine tools	People, software
Reservoirs	Capacitors, batteries	Stock, on-line buffers	Filing cabinets, databases
Taps	Switches, transistors	Orders, kanbans	Database search request

1.3.3.2 Limitations of the operational definitions of information

The definitions of information given above can be seen as limiting since they can lead to the conclusion that information can be transported and processed automatically just as any other material resource in an organization. However, as has been pointed out, one of the crucial points to bear in mind at this point is that people add value to information and that people-to-people always seems to be the preferred transport mode of information in an organization. This is a serious limitation to the analogy given above, where in the case of material or energy flows, transportation does not and cannot change the nature of the flow. In the case of information, however, people are capable of changing the physical nature of the information flow by translating it into another language for instance, or writing it down on paper, or adding value to it generally by using their knowledge.

Three other important properties of information are not represented in the operational definitions of information:

- the negligible costs of transfer and distribution;
- the free disposal of information;
- the high cost of screening.

These properties are, according to some authors, the causes of the information overload since transmitters are encouraged to be indiscriminate in broadcasting their output.

1.3.4 Use of information

So far, the issue of how best to describe information in order to get a global and yet useful characterization of information has been addressed both in general terms and also narrowed down to the information used in manufacturing. It has been shown that looking at information as a resource could prove useful in the study of MIDS. The description proposed does not, however, help understand why information is necessary in a manufacturing system.

Most analyses and definitions of manufacturing information are based around the fact that there are three basic uses for information, which can be summarized as:

- information used for decision making (e.g., market survey);
- information used to control a process (e.g., production plan);
- information used for pure consumption (e.g., reading a newspaper) (Ouwersloot, Nijkamp and Rietveld, 1991).

This classification does not take into account the educational value of some information, and in fact a further use of information should be added:

■ information used to increase skills or knowledge.

The usability of a piece of information is obviously an important property of that piece of information. It is, in fact, when value aspects of information systems are being looked at, probably the most important property of that piece of information. It is determined by comparing the reason for that piece of information's presence in the system with its actual use and role:

■ whether the information is wanted in the system;
■ whether the objective of the information is being achieved.

The way information is used is addressed in *semiotics*. This considers all aspects of communication and therefore information. Various semiotic levels are defined, as shown in Table 1.4.

It can be argued that semiotics is a powerful tool for analysing an information system because it is independent of any particular technology. The message analysed in Table 1.4 could be, for instance, either a computer-to-computer message sent across a local area network (LAN), a message sent to a foreman using a paper form or a spoken instruction. But it is obvious that semiotics is very powerful in that it addresses all aspects of information looked at so far in a unified manner.

1.3.5 Specificity of manufacturing information

The most common way of analysing manufacturing information is by using an information intensity matrix as shown in Figure 1.7a

Table 1.4 *Semiotic analysis of a manufacturing control message*

Semiotic level	Interpretation
Organizational context	Need to produce a part
Pragmatics	Reason for sending the message
Semantics	Information content of the message content
Syntactics	Protocol, language
Empirics	Noise
Physical world	Local area network, paper

(Porter and Millar, 1985). This matrix is founded on the fact that products can be more or less information intensive and that processes required to produce such products are also more or less information intensive. In the example of oil refining given in Figure 1.7a, the processes, known as a *value chain*, are clearly very information intensive, requiring complex engineering and control. The product itself, on the other hand is of a low information intensity as it does not contain usable information.

The examples given in the matrix in Figure 1.7a show that it would be impossible to position manufacturing in general on this matrix for the simple reason that it will depend on the type of products being produced and the processes used. It has also been seen above that the concept of *pragmatics* (the *why* of information) could help to address the issue of the performance of information systems. The concept of value also seems to be very relevant. If a product is highly information intensive, the information needed to reach that high level of intensity would have had to have been conferred through the value chain, unless the raw materials are themselves highly information intensive, which is by their nature rare. In order for this process to be efficient, the product would need to acquire its information intensity using the lowest amount of resources possible. A similar matrix can therefore be defined in order to represent MIDS value, which is shown in Figure 1.7b. The figure shows that the MIDS of the manufacturing enterprise will be more or less critical as far as the value of the product is concerned, according to the position in the matrix. For instance, an information-intensive MIDS will generally bring value to a company

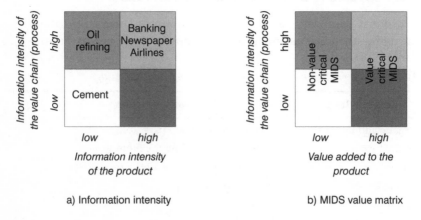

a) Information intensity   b) MIDS value matrix

Figure 1.7 *Information intensity and value matrices*

if it adds high value to the product. In this case, the MIDS will also be critical for the company, as far as the value of the product is concerned, i.e., value-critical MIDS.

The value added to the products in the form of manufacturing information will therefore need to be considered as a major performance criterion of MIDS. It seems, however, that such values can be difficult to quantify at this stage, and manufacturing information needs to be further examined in order to classify it in such a way that will allow the determination of the value added by each information element.

Manufacturing information is, by nature, complex and heterogenous and its main characteristics can be outlined as:

- complex data types (vectors, matrices, sets, etc.);
- composite and complex objects;
- multiple versions of designs;
- hierarchies for data structures;
- attributes that draw values from alternative domains;
- recursive definition of data objects;
- temporal, positional and procedural relationships.

Some of these characteristics can be immediately related to value. That complex data types are necessary is due to the fact that conferring a value to a manufactured product involves engineering. Such an approach to describing the nature of manufacturing data is useful in order to set the minimum requirements for an MIDS, but it tends to draw the attention away from value to functionality.

1.3.6 The dual nature of manufacturing information

It has been shown so far that manufacturing information has to be considered as an asset of the manufacturing enterprise. It could, however, be argued that in some cases manufacturing information is only present in the production system as a means to get a product manufactured, but not directly adding value to the product. Producing, transporting and storing such manufacturing information are, in effect, activities that do not in themselves add value to the product, but merely ensure that sufficient value is added to the raw materials that arrive in the enterprise in order for the products that come out to be marketable at a profit.

Adding value to raw materials, however, involves adding information to them, which, to use a physical concept of information, is equivalent to adding order to the raw materials. In this particular view of manufacturing information, generating the information necessary to add value to the product is a value-adding activity. It could also be argued that storing this information, transforming it or transporting it are also value-adding activities since we are talking about an asset of the enterprise.

At this stage, a way needs to be found to differentiate between the various types of manufacturing information, and their various properties, which are either necessary or unnecessary to the achievement of the prime goal of the enterprise. This leads to the introduction of the concept of the *critical characteristics of manufacturing information* (CCMI). These characteristics need to be defined at a pragmatic level, in other words, related to the reason why the information element is needed with regards to the overall aims and objectives of the enterprise. This allows us to define five CCMIs, which are not mutually exclusive. Three of those can be defined at a pragmatic level:

- product value criticality;
- operations criticality;
- personnel criticality.

As with any other operational resource, there are two other variables that can be used to characterize information and which will determine the way the information should be handled: (i) the time dimension, and (ii) whether or not the right information is there, which can be taken into account by introducing an accuracy dimension. Two further CCMIs will therefore be defined at an operational level:

- time criticality;
- accuracy criticality.

1.3.6.1 Product value

As mentioned earlier, a number of manufacturing information elements are essential for adding value to products. A typical example of this would be design information without which the product would simply not exist, or exist without added value. It can be argued that each element of manufacturing information needs to be characterized by its degree of value addition to the company's products.

The definition of a value-critical information element can therefore be given as an information element that contributes directly to the value of the manufactured product by:

- contributing to its design;
- ensuring its price;
- ensuring its quality.

1.3.6.2 Operations

Some information elements will be essential to the smooth running of operations in the enterprise, just as transport operations in a manufacturing system are essential to the flow of materials. These information elements do not, however, add value to the products. They can be commands that direct a part to a machining centre, or shop-floor data about the status of an assembly. For instance, stock data could also be considered to be *operations-critical*, but not value-adding data. These operations-critical information elements are typically those of which the generation, transport and storage can easily be automated.

An operations-critical piece of information is therefore defined as an information element that contributes directly to making the product either:

- quickly; or
- on time.

1.3.6.3 Personnel

Some manufacturing information can be seen as *personnel-critical*. Although such information is not strictly necessary for general manufacturing operations, it is necessary to ensure smooth running of the organization in the future. A chart on the wall that displays productivity figures or company profits can, for example, be seen as necessary to create the right feel-good factor on the shop floor.

The distinction between personnel-critical and operations-critical information is difficult to make since the wall charts described earlier can be seen as some form of control over the workforce. The main differentiating factor here is probably timescale. Whereas operation-critical information is likely to be created, transported, used and stored along the same timescale as

the products going through the system, personnel-critical information as defined here should be seen as evolving along a much
longer timescale.

A definition of a personnel-critical piece of information can be
given at this stage as an information element that contributes
directly to maintaining or improving the level of:

- knowledge;
- morale of the company's employees.

1.3.6.4 Time

Being generated on time or arriving on time is more important for
operational information than it is for other types of manufacturing
information such as design concepts. The time criticality of a piece
of information will also vary according to the form it takes. A
customer order arriving through the postal service could be seen as
time-critical whereas the same order transmitted by electronic data
interchange (EDI) will have lost its time criticality since it is less
likely to arrive later than the posted order. Time criticality will,
therefore, depend on the pragmatic nature of the information
element considered, the timescale operated by the receiver and the
timescale operated by the sender. It can therefore be measured or
estimated and given as a set of numbers linked to events in the life
cycle of the information element.

A definition can be given for the criticality of a time-critical information element as a set of numbers defining the specific time
periods or the specific points in time during which an information
element needs to be generated, transported and/or used.

1.3.6.5 Accuracy

It can be argued that all information is accuracy-critical insofar as
maximum accuracy is always desirable. However, there are generally
special treatments that are required for information elements on
the grounds of accuracy. For example, care must be taken with the
accuracy of dimensions in a design for a product that is to go into
production, but the same care is not necessary for a concept
drawing. In a similar fashion, the same level of accuracy is not
required for sales forecasts that look three years ahead than for forecasts that look at the following week and on which the production

plan will be based. Although desirable, the accuracy of the three-year plan could, for instance, involve expensive market surveys and computing intensive data manipulations that might not be justified on the grounds of the level of accuracy required.

Just as in the case of time criticality, accuracy criticality will depend on the pragmatic nature of the information element considered, but is not as easily measurable in all instances. For numeric data, a maximum allowable error can be given, but the case is less clear for higher-level information.

A definition can be given for the accuracy criticality of an information element as a set of measures or other indicators quantifying the accuracy with which an information element needs to be generated, transported or used.

1.3.6.6 Example of CCMI analysis

The information flows within the major subsystems of BCIM are shown in Figure 1.2. These can be taken one by one and analysed for their critical characteristics. An order, for instance, which is an information element going from the sales front end to the production planning and control subsystem, is:

- not value-critical, since it contributes neither to the design of the product, nor to its price, nor to its quality;
- operations-critical, since the arrival of the order and the message contents both contribute to the product being made on time;
- not personnel-critical as such since, at least in the case of BCIM, a single order would not contribute directly to the knowledge or the morale of the employees;
- time-critical, since the order needs to be transmitted as soon as possible to production planning and control from sales front end;
- accuracy-critical, since it is important that the message transmitted gives the right information in terms of quantity, type of product and other required design specifications.

Table 1.5 shows the various CCMIs for the information elements in BCIM. Criticality is represented by '+' whilst non-criticality is represented by '–'. In the case of time criticality and accuracy criticality, '+' indicates high criticality whilst '–' indicates low criticality. It is interesting to point out that this table shows that, in the case of BCIM, over half of the information elements identified are value-critical. A much higher proportion is, however, operations-critical.

Table 1.5 *Critical characteristics of the manufacturing information in BCIM*

Information element	Information Flow		Critical Characteristic of the Manufacturing Information				
	From sub-system	To sub-system	Value	Oper-ations	Person-nel	Time	Accura-cy
Order	Sales	PPC	−	+	−	+	+
Schedules	PPC	Cell	−	+	−	+	+
NC codes	CAD	Cell	+	+	−	−	+
Process plans	CAD	PPC	+	+	+	−	+
Design Specs	Sales	CAD	+	−	−	−	−
Tool Offsets	CAQ	Cell	+	+	−	+	+
CMM data	Cell	CAQ	−	+	−	+	+
Quality reports	CAQ	Sales	−	+	+	−	−
Quality reports	CAQ	CAD	+	+	−	−	−

1.3.7 Performance maximization rules

CCMI allows a classification of manufacturing information according to its effect on critical performance indicators in the company. It is now desirable to determine a set of rules or basic guidelines that would help to look at the performance of manufacturing information systems. A rule of thumb would be that operations-critical information needs to be minimized, since it does not in itself contribute to the value of the products, and in that sense to the basic performance of the company. This is, in some ways, the case when just-in-time (JIT) is introduced as a replacement of MRP II for scheduling, and when it is often argued that complex MRP II information systems are unnecessarily bureaucratic when simpler systems can work.

Minimizing operations-critical information means that all information not strictly necessary for the smooth operation of the systems should be eliminated, if (and only if) the information is not also value-critical. Also, from an intuitive point of view, value-critical information should be maximized. In other words, every opportunity should be considered for value to be further added to this information.

The various qualities that should be looked for in information will therefore vary according to the critical characteristics of that particular information, and thus will be examined in more detail below in order to determine performance maximization rules for each type of critical characteristic.

As a starting point, the information quality measures published in the literature can be used (DeLone and McLean, 1992). With these, a set of performance maximization rules for each information element can be built. The information quality criterion is considered first. The measures that are proposed for information quality range from importance to freedom from bias, and are not all applicable to every information element, depending on their critical characteristic. For instance, time-critical information does not need to be important, relevant or useful, but it will have to be usable if it is not to lose its time criticality. Each of the information quality criteria can therefore be analysed as to their potential for performance maximization of MIDS dealing with information elements. A summary of the analysis is shown in Table 1.6. One of the main difficulties in establishing the general rules presented in Table 1.6 is that the CCMIs have been defined as being non-mutually exclusive. This makes it difficult to establish general rules about information quality. The table here is nonetheless potentially useful and can be used as a basis for the establishment of rules specific to the systems analysed.

Once various information flows and elements have been identified in the system and characterized by their CCMI, they should be checked against the qualities that should be maximized. For example, value-critical information should be checked for its relevance, usefulness, usability, readability, clarity, content, conciseness and reliability. Operations-critical information should be checked for the same qualities as well as format and uniqueness. If all these qualities are present and maximized, it does not prove that the best information system has been achieved, but it means that under the current architecture, the information system is likely to be successful as far as information quality is concerned. There will obviously be instances where one characteristic of the manufacturing information will require one quality to be maximized, whereas other constraints will go against this maximization. In this case, a trade-off will be necessary and will have to be decided on a case-by-case basis.

Table 1.6 *Performance maximization rules*

Information Quality measure	Critical characteristic of the manufacturing information				
	Value	**Operations**	**Personnel**	**Time**	**Accuracy**
	(I = 'intrinsic' '+' = to be maximized 'NA' = quality not applicable and '=' = no general rule)				
Importance	I	I	I	NA	NA
Relevance	+	+	+	NA	NA
Usefulness	+	+	+	NA	NA
Informativeness	=	=	+	NA	NA
Usability	+	+	+	+	NA/+
Readability	+	+	+	+	+
Clarity	+	+	+	+	+
Format	=	+	=	+	+
Appearance	=	=	+	=	=
Content	+	+	+	+	+
Accuracy	=	=	=	=	I
Precision	=	=	=	NA	NA
Conciseness	+	+	+	+	=
Sufficiency	NA	NA	+	NA	NA
Completeness	NA	NA	+	NA	NA
Reliability	+	+	+	NA	NA
Currency	+	+	+	NA	NA
Timeliness	+/=	+	+	I	NA
Uniqueness	+/=	+/=	+/=	=	+/=
Comparability	NA	NA	NA	NA	NA
Quantitativeness	NA	NA	NA	NA	NA
Freedom from bias	+	+	+	+	+

1.4 MIDS taxonomy

1.4.1 Concept of MIDS

So far, the importance of manufacturing information in the context of modern manufacturing has been highlighted, particularly computer integrated production. No definition has, however, been applied to MIDS. The issue has received very little attention in the manufacturing engineering literature, which tends to be biased either towards conceptual descriptions of manufacturing systems that concentrate on operational systems, on low-level control, or on production management philosophies.

In the same way as there is no one definition for CIM, there will be no one definition for MIDS. However, the descriptions below are generally used in manufacturing planning and will be useful in later chapters for addressing issues pertaining to the management of MIDS.

In general, manufacturing control systems are seen as implementation of two major activities, namely *factory co-ordination* and *production activity control*. This analysis is based on a functional decomposition of the activities (tasks) involved in factory management (Gupta and Biegel, 1991).

At the factory level, production management is concerned with factory co-ordination, which can be further subdivided into two functions. One is concerned with the design of the production environment and the other with overall factory control. In this architecture, control activities comprise five sub-functions (Figure 1.8):

- scheduler;
- dispatcher;
- monitor;
- mover;
- producer.

This decomposition is then translated into a generic architecture for MIDS. At this point, describing MIDS is just a matter of describing the implementation of the architecture proposed. For instance, an implementation of a factory level scheduler would be a material requirements planning (MRP) system and a factory activity monitor could take the form of a management information system (MIS).

The mapping of the generic conceptual architecture to physical systems is, however, not necessarily so easy, insofar as many of the tasks required by factory co-ordination are, in fact, carried out by humans. Furthermore, some information systems will perform only parts of some activities or will overlap activities. This is, in fact, the case for the MRP systems, which on one hand deal with bills of material that would fall under the production environment design task, and on the other hand production requirements, which would fall under the factory control task.

The model proposed here does not include activities concerned with the design of products, which must, of course, be linked with production management activities, but the analysis can be generalized

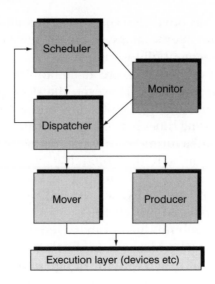

Figure 1.8 *MIDS at factory level*

to include design. It can also be argued that MIDS need to provide information for the strategic management of the enterprise. The two functions therefore need to be added to the model in order to describe MIDS accurately. The proposed model is, however, interesting since it is based on the observation that in practice all the activities that take place within a manufacturing information system are concerned with co-ordination and design.

1.4.2 CIM and MIDS

It has been argued in Section 1.1.2 that information is the main vehicle of integration, and a number of companies are trying to implement CIM as a way to improve the efficiency of their operations, both in the short and long term. CIM is often confused with automated cellular manufacturing or simply the introduction of advanced manufacturing technology. In reality, this type of definition is very limiting, since it focuses on the computer and manufacturing aspects of the definition, whereas the integration aspect is that which leads to efficiency savings and improvements in lead times, for instance by reducing paperwork and facilitating concurrent engineering.

CIM is often described in the literature with the aid of the CIM pyramid, as shown in Figure 1.9. The pyramid generally shows different levels of control in a typical manufacturing organization, and the exchange of information between or within levels. Vertical integration refers to the exchange of information across levels of the pyramid. For instance, the linking of CAD at the design level with CAM at the shop floor level through the automated generation and uploading of numerical codes for the machine tools is a typical example of vertical integration. Horizontal integration refers to the exchange of information within levels. For instance, the linking of machine tools to a PC controller at shop floor level could be seen as some form of horizontal integration.

Other CIM models are based on the fact that in reality the emphasis is not so much on the hierarchy of functional areas but on the hierarchy of information technology support. This is the case of the IBM model, as shown in Figure 1.10 (IBM, 1990).

In the IBM model, the emphasis is on the idea that a typical CIM environment can be divided up into six functional areas, which work together (vertical integration):

- production operation;
- production planning;
- engineering and technical computing;
- marketing and product support;
- business management;
- distribution and logistics.

Three services are needed to support these manufacturing functions (horizontal integration):

- application development support;
- decision support;
- administration support.

Figure 1.9 *The CIM pyramid*

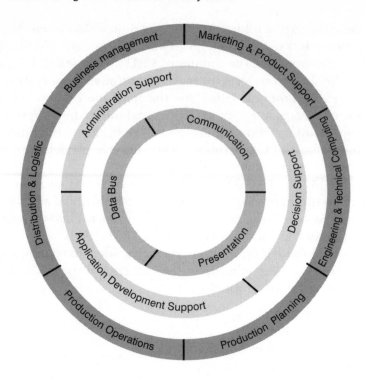

Figure 1.10 *The IBM CIM wheel*

These three services depend on three aspects of information technology:

- data bus;
- presentation;
- communication.

All these parts of information technology work together (horizontal integration again) to underpin the support services and will be presented later in this book.

1.5 Conclusion

This chapter sets the scene for the rest of the book. Information systems and hence manufacturing information systems should not only be seen as a collection of hardware and software, but, equally important, as the people and organizations behind these technical systems, who in turn should be taken into account when analysing performance, simply because they are integral parts of the information system.

Various definitions of manufacturing information have been given and they show that no definition is appropriate on its own. Additionally, the examination of properties of manufacturing information shows that manufacturing information can be classified according to five critical characteristics that are either pragmatic (value, operations and personnel criticality), semantic (personnel and accuracy criticality) or empirical (accuracy and time criticality). In determining which qualities should be maximized for each type, these characteristics help understand how information flows will have an influence on the performance of the MIDS and by extension the company in general.

A description of MIDS has been proposed, which is based on a functional decomposition. This is by no means the only possible description that can be proposed, but it is sufficient to show that they are complex socio-technical groupings of functions involving manufacturing technology, people, information and material flows. This heterogeneous grouping of functions implies that MIDS are complex systems, the impact of which will be difficult to assess, as will be shown later in the module.

1.6 References

Broomhead P, Grieve R, Schmid F (1989). The teaching of CIM in a university environment, *Proc. 6th IMC Conf. on Advanced Manufacturing Technology*, Dublin City University, pp. 619–635.

DeLone W.H, McLean E.R (1992). Information systems success: the quest for the dependant variable, *Information Systems Research*, 3 (1) pp. 60–95.

Gershwin S, Stanley B (1994). *Manufacturing Systems Engineering*, Prentice Hall, Harlow.

Gupta U, Biegel J (1991). ManIS: manufacturing information systems, *Computers in Industrial Engineering*, 21 (1) pp. 235–239.

IBM (1990). *CIM Solutions for Midsized Companies*, IBM corporation, US Marketing and Services, Dept. ZVO, 1133 Westchester Avenue, White Plains, NY 10604, USA.

Jayaratna N (1994). *Understanding and Evaluating Methodologies: NIMSAD, a Systematic Framework*, McGraw-Hill, London.

Liebenau J (1990). *Understanding Information: an Introduction*, Macmillan, London.

Mitchell F (1991). *CIM Systems: an Introduction to Computer-integrated Manufacturing*, Prentice Hall, Harlow.

Ouwersloot H, Nijkamp P, Rietveld P (1991). Economic aspects of information and communication: some considerations, *Information and Software Technology*, 33 (3) pp. 171–180.

Porter M.E, Millar V.E (1985). How information gives you a competitive advantage, *Harvard Business Review*, pp. 149–160.

Sartori L (1988). *Manufacturing Information Systems*, Addison Wesley, Harlow.

Shannon C (1949). *The Mathematical Theory of Communication*, University of Illinois Press, London.

Stonier T (1990). *Information and the Internal Structure of the Uuniverse: an Exploration into Information Physics*, Springer-Verlag, London.

1.7 Further reading

Galliers R (1992). *Information Systems Research: Issues, Methods and Practical Guidelines*, Blackwell Science, Oxford.

Steudel H (1992). *Manufacturing in the Nineties: How to Become a Mean, Lean, World-class Competitor*, Chapman & Hall, London.

Willcocks L (1994). *Information Management: the Evolution of Information Systems*, Chapman & Hall, London.

2

Manufacturing Databases

2.1 Database systems

2.1.1 Database concept

A *database* is defined as a collection of stored operational data used by the application systems of some particular enterprise. For instance, a simple filing cabinet or a file system on a computer hard disk can be considered to be databases.

The internal structure of a database is such that data are stored to represent the physical entities[1] of the system, and the relationships that link the entities. These relationships may be represented by pointers, or physical adjacency, and they may be either simple one-to-one relationships, or complex two or three-way relationships. The physical entities possess attributes that describe the features of the entity, each attribute having a domain, or range of permitted values. The way in which the entities, attributes and relationships are represented in the database is dependent on the particular database model used.

The simplest form of computer databases are file systems, where the data are stored in various files, and the access is made through third generation programming languages such as BASIC, COBOL or FORTRAN. Although this method is still used when data are kept in spreadsheets, for instance, keeping data in a file system is

---

[1] Entity: something having real or distinct existence; a thing, esp. when considered as independent of other things; existence or being the essence or real nature (Oxford Dictionary).

equivalent to keeping a computerized list of records. This method has a very large number of disadvantages:

- the whole database is searched for a key when the data are needed. This is more or less equivalent to a search/replace procedure in a word processor, i.e., slow;
- producing selective reports is extremely difficult;
- experts are needed to write access programs to the data;
- the data structure is structurally dependent. Any modification of the structure of the data file, even its location, will require changes to be made in all the files and programs that use the modified file.

2.1.2 The evolution of database systems

2.1.2.1 Records and cross-referencing

The first computer database systems for use in manufacturing were of a form that was similar in structure to the paper-based filing cabinet, methods they were designed to replace. As an example, one of the early paper-based methods could apply to a part known by a particular name, commonly used in a company. Keeping records of parts by using names presents problems, particularly when there may be similar or even the same names used for other parts or there may be a generic group where all the parts have the same name but are geometrically different. The usual way to overcome this problem is to assign a unique part number to each part. It is now possible to keep a record of all the parts used in the company using the part number as the primary method of locating data about the part. The contents of the record consist of all the data concerning the part that is of relevance to the company. Providing the part number is known, it is relatively straightforward to obtain information about the part from the database.

Unfortunately, information is often required about a part where the part number is unknown but some other identifying characteristic is known. An example of this may be when the part has been supplied by a vendor, and the vendor's part number is stored in the record of the part. If at some stage the company wished to know the price paid for the part the last time it was ordered or the size of the order, it might be more appropriate to search for the information required by using the vendor's part number rather than the company's part number. If the record location is based solely on the company's part number this would mean that every record would have to be searched to extract the information required. This would not be a very efficient method of proceeding.

Cross-referencing methods have been developed to overcome this problem. Separate lists are maintained that can be ordered by part number, names in alphabetical order or other descriptions of use to the company. The lists take the form of the simple tables shown below where the part number and the part name are cross-referenced. If the part name is known, the part number can be determined and the information about the part obtained. An example of cross-referencing is shown in Tables 2.1 and 2.2.

These records and cross-reference lists have to be maintained and could require significant updating effort when new parts and processes are introduced. For a company with relatively few parts, this system can perform efficiently. However, even relatively modest companies are faced with increasing problems as the number of parts, sub-assemblies and processes grow. This method also relies on the fact that the types of queries that can be expected are known at the time the system is first designed. For example, if the vendor part number was seen to be an identifying characteristic of importance, a cross-reference list of some type could have been created at the design stage linking that number back to the

Table 2.1 *A cross-reference list: part number to part name*

PART NUMBER	PART NAME
A1234	CIRCLIP
B5678	SHAFT
J3456	LOCKNUT
L9541	ELECTRIC MOTOR

Table 2.2 *A cross-reference list: part name to part number*

PART NAME	PART NUMBER
CIRCLIP	A1234
ELECTRIC MOTOR	L9541
LOCKNUT	J3456
SHAFT	B5078

company part number. If not, the creation of a list, as the example shows in Table 2.3, after the system has been running for some time would be a major undertaking.

Using this method for the example described above, the cross-reference list based on the vendor's part numbers would first be accessed and from it the company part number determined. This could then allow the record system based on the company part number as the prime location feature to be interrogated. Using paper-based systems for record keeping is not straightforward and can become complicated, labour intensive and error prone.

2.1.2.2 Sort and report file management system

The early computer-based database methods solved the problem of record and cross-referencing data handling by using what is known in general as *sort and report file management*. The records of the parts used in the company, employees, sub-assemblies or products, would be kept in some form of logical order such as by part number. Each record would contain all the relevant information required by the company about the part such as part name, part description, classification code, cost, vendor and so on. The speed of the computers meant that a cross-reference system would not be required even though the records may have been ordered by part number. If, for example, the parts supplied by a particular vendor were required, then the whole database would be searched and eventually the required parts sorted and reported. This was an obvious improvement over the paper-based methods but even so the

Table 2.3 *Vendor part list*

VENDOR	INTERNAL PART NAME	VENDOR PART No	INTERNAL PART No
ABC	CIRCLIP	2256	A1234
	SHAFT	Q734	J3456
	...	...	...
	...	...	...
XYZ	ELECTRIC	AB24	L9541
	MOTOR SHAFT	CD72	B5678
	...	...	...
	...	...	...

systems were not readily accessible to non-experts, and reports on, say, stock levels or other matters of interest to the company would be generated on only a monthly or a two-weekly basis. Information would therefore be old by today's standards, although at the time the improvements over the previous methods were impressive.

With relatively small amounts of data, the sort and report method proved satisfactory, but as greater demands were put upon database systems to accommodate more data (hundreds of millions of records may exist in some applications) the need to search the whole database every time provided a limitation based on the time required for the search if nothing else. Improvements in the design of database management systems led to the introduction of cross-referencing and linking through the use of tabular structures. This improved speed and efficiency and also began to satisfy the demand for an instant access to the database by non-experts using networks of computer terminals.

2.1.3 Database management systems

More advanced storage structures than the ones presented above are supplemented by suites of computer programmes that perform data access and modification. These two components form the database management system (DBMS). Figure 2.1 shows the structure of a DBMS and the way it provides an interface between the user and the manipulation of data (Burch and Grudnitski, 1989).

Perhaps the greatest difference between a DBMS and traditional file organization is that the DBMS separates the logical and the physical views of the data, relieving the programmer or end-user from the task of understanding where and how data are actually stored. Hence, a DBMS combines the hardware and software for creating, operating, and maintaining a database. However, when describing the functions of a DBMS, emphasis has to be placed on the functions concerned with the data, such as in the list of functions summarized in Table 2.4 (Oxborrow, 1990).

The DBMS essentially provides the interface between the users with their application programs, and the database itself. There are 10 functions that a genuine DBMS needs to be able to perform (Rothwell, 1993). These are:

Function 1:   a DBMS must have the ability to store, retrieve and update data in the database. This is the fundamental function of a DBMS.

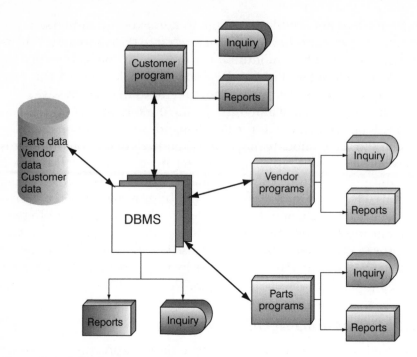

Figure 2.1 *A database management system*

Function 2:   a DBMS must ensure that, when an update operation is performed, either the entire operation is carried out successfully or that none of it is done. If the update operation is interrupted before it is complete, the system must be able to roll back to the unchanged state and inform the user that the update has failed to ensure data accuracy.

Function 3:   a DBMS must ensure that updates are handled correctly even though many users may be updating the database simultaneously. Again, to ensure data accuracy, the DBMS must either be capable of handling simultaneous inputs or only allow write access to a file to be open to one user at a time.

Function 4:   a DBMS, if damaged by an accident, must have the means of preserving the data in an unblemished state. Power failure is the most common type of accidents.

Function 5:   a DBMS must allow its users to access communications software. This is because access to databases is

Table 2.4 *DBMS functions*

Facility	Comments
1. Data description	
● record, relation, etc., description	
● structured data description	● Not relational systems
● view (sub-schema) description	● Not always provided
2. Data input/edit/update	
● file-based (large databases)	● Important for large databases
● interactive/form-based (micro DBMS users)	● Important for micro DBMS users
3. Data retrieval	
● interactive query language	● Generally non-procedural and at various levels
● other data retrieval / manipulation language	● May be record-at-a-time/may be extension of non-procedural query language
4. General-purpose data manipulation	
● application programming facilities via host or self-contained language	
5. Data presentation	
● tabular display of results	
● report facilities	● Generally quite sophisticated in miniframe DBMSs
● graphical display	
6. Data protection	
● passwords, authorization rules	● For access authorization
● back-up copy, transaction logs (recovery after failure of some sort)	● For recovery after some sort of failure
● validation rules (database integrity)	● Database integrity assurance
7. Data administration	
● integrated data dictionary	● Not always provided
● usage monitoring	
● physical organization facilities	● For database tuning

usually from remote terminals not directly connected to the hard disk on which the data is stored.

Function 6: a DBMS should provide a dictionary or catalogue that lists all the data items accessible to the users. This aspect, normally called the *data dictionary*, is necessary to increase user-friendliness, and consequently the efficiency, of the system by detailing the information that is accessible.

Function 7: a DBMS must ensure that only authorized users can access the database. Unauthorized access is highly undesirable as the database may contain sensitive information or the data may be corrupted.

Function 8: a DBMS must insist that both inputs of data and changes to such data follow certain constraints so as to ensure the integrity of the database.

Function 9: a DBMS must be able through the actual structure of the database to ensure the independence of programs. This is necessary so that it is possible to change the structure of the database without having to modify the programs that use it.

Function 10: a DBMS should provide a set of utility functions. This is an umbrella function designed to cover all improvements to DBMSs and includes features such as providing statistics on patterns of database usage.

2.2 Database models

There are three established data models used in databases today:

- the hierarchical model;
- the network model;
- the relational model.

Research has suggested the use of *object-oriented programming* to provide efficient data models for manufacturing automation. Object-oriented databases are now available commercially, although at present their use is restricted to research laboratories.

Historically, the hierarchical model and the network model were developed around the same time, during the 1960s, while the relational model was not proposed until several years later. Both the network and hierarchical models are tightly linked to the application chosen, while the relational model is more flexible, and not constrained in this way.

2.2.1 Hierarchical databases

The structure described above provides a means of storing data that is of use to the company but no attempt is made to look at the relationships between the individual elements of the data. Typically, the company may be keeping records of the assembly of a finished product that contain data relating to cost, number in stock, the part elements that go to make up the assembly, the engineering drawing codes and other manufacturing information. This information can be considered to be in the form of subsets, such as the list of parts used in an assembly being a subset within the total amount of data on the assembly. This tree-like form of representing data leads to the implementation of hierarchical structures that are difficult to operate using paper-based methods but can be run using computers.

The hierarchical database model uses records and links to represent data and relationships and the records are organized as a collection of tree structures. Four attributes characterize the hierarchical data model:

■ the model always starts with a root node. It is the only node in the tree that does not have a parent;
■ every node consists of a number of attributes describing the entity that is exemplified at that node;
■ each node, excluding the root node, can have only one parent, but each parent node can have many children;
■ each node is reached or retrieved by passing through all the preceding nodes (Rothwell, 1993).

Designing a database management system to operate efficiently over the hierarchies of data that exist in many companies is not a trivial task but, once implemented, allows access to data with reasonable ease. For example, if information is required about all the parts that go to make up an assembly, then the product code would be located first, and then using the parts path of the hierarchy retrieve all the parts data. Figure 2.2 show this schematically using the example of a car, although in reality a complete car is too complex an assembly to be held in this manner.

Although a number of different methods of accessing data through hierarchical databases have been developed, the basic methodology

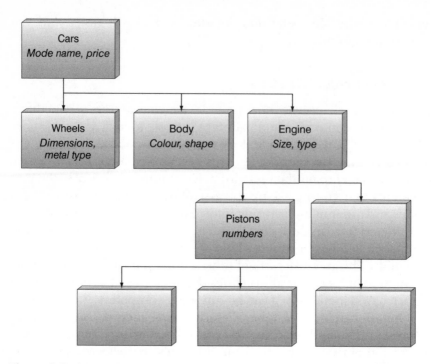

Figure 2.2 *A tree structure*

is pre-order transversal, or first son, next brother, which is shown in Figure 2.3. This can be easily done because the data is stored physically using pointers for each record that point to the record's father, its first son and its first brother. It can be seen from Figure 2.3 that accessing data at the left hand of the tree structure is a slower process then accessing data at the right.

Although the introduction of hierarchical databases has brought benefits over the previous filing methods, there are some drawbacks to the method. It is relatively easy to locate all the parts in an assembly by locating the relevant hierarchy structure and moving along the required path until all the required data is obtained. A hierarchical structure, however, implies that there is a separate structure of each of the assemblies produced by the company but it may be the case that the same part is used in more than one assembly. This means that the same part record would have to be maintained in more than one hierarchy. Duplicate copies of the part record will now appear in the database, which leads to redundant data and requires a major database search when determining which

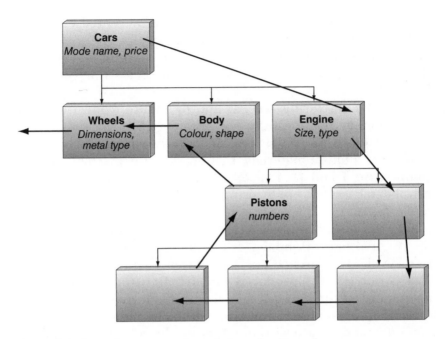

Figure 2.3 *Pre-order transversal search through a hierarchy*

products are affected if a part is in some way changed. It is important to note that the representation of many-to-many links is not possible within the framework of the hierarchical structure, unless records are replicated.

The problem of data inconsistency can be overcome by introducing *virtual records*. Virtual records take the place of the replicated record, and contain only a pointer to the actual record. There is still a waste of space, however, as the virtual records require as much storage space as an ordinary record. An example of a database systems based on the hierarchical model is the IMS (information management system) produced by the IBM corporation.

2.2.2 Network databases

To some extent, the problem of cross-hierarchy relationships was overcome by the introduction of network database models, which endeavoured to allow such relationships. In essence, the network database model appears to be a variation of the hierarchical data model. In practice, data can be translated from hierarchical to network and vice versa to optimize processing speed and convenience.

In the network model, the entities are represented by *records*, and the relationships between the data entities are represented by *links*. The records are organized in the form of a graph with the links forming associations between records, as shown in Figure 2.4.

Most manufacturing records are, by nature, concerned with many-to-many relationships. For instance, the same component can be used in more that one end product (e.g., if Figure 2.4 represents a bill of materials, **L** appears in the assembly of both components **I** and **J**). This could not be represented in hierarchical databases, except by using virtual records, hence the many replications.

Since its inception, however, changes have been proposed to the original network model. For example, in the late 1960s the CODASYL (Conference on Data Systems Language) Database Task Group (DBTG) produced a model after studying the implementations of the network model existing at that time, in order to develop an industry standard. Although their first report in 1969 received a great deal of criticism, the recommendations in the revised report in 1971 have since become the yardstick for vocabulary within the network database model world. One important contribution from CODASYL was their definition of the logical structure of a database by inventing the word *schema*. The schema is the overall logical view of a database, and has four components (Rothwell, 1993):

The *schema entry* gives the name of the schema and its author, date of completion and any other introductory remarks.

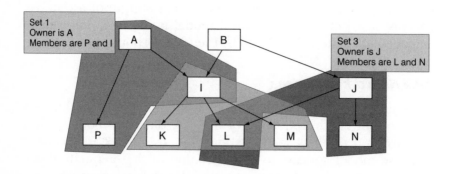

Figure 2.4 *Data structure in a network database*

Example:

SCHEMA NAME IS SCRAPE AND BLOW LTD

Schema compiled by Xanadu Analysts Completed January 1991

The *area* (or *real*) entry identifies the area of the database in which tables of data are stored, and the details of records that subsequently appear will be appended to the appropriate area.

Example:

AREA NAME IS AREA_DISTRIBUTION

The *record* entry declares records or tables that exist in the schema under two sub-entries:

- *record sub-entries* give the name of each table, how it can be accessed, and whereabouts it may reside;
- *data sub-entries* list the attributes of each column within the table.

Example:

RECORD NAME IS COURSES

LOCATION MODE IS CALC

USING COURSE_NUMBER WITHIN AREA_DISTRIBUTION

02 COURSE_NUMBER	PIC 9(3)
02 NAME	PIC X(20)
02 DEPARTMENT	PIC X(20)

The *set entry* declares the associations among the various types of record, for example, as we know a record (in a network database) can have more than one parent, this is the part of the schema which would define this.

Example:

SET NAME IS SYMPHONIES **OWNER IS WORK** **MEMBER IS PERFORMANCE**

In this example, a relationship is created between **WORK** and **PERFORMANCE**, by defining a set containing the parent (**OWNER**) and the child (**MEMBER**).

As well as the schema, it is also possible to define a *subschema*. This is similar to the schema but less thorough and relating to only one part of the database. These are used in application programs where use of the complete schema is unnecessary, and would therefore waste processing power.

The DBTG model constrains relationships, or links, to be of cardinality one-to-one or many-to-one, thus relationships linking one or more entities to many others cannot be represented explicitly with this model. Each linked pair of records is known as a *DBTG-set*, with one record the owner of the set, and the other the member of the set. Commercial examples of this database model exist today, for example IDS11 from Honeywell and DMS 1100 from Sperry Univac.

Both hierarchical and network databases are heavily application-dependent and once developed require a major effort to change. Providing that the initial design is good and all eventualities are covered in terms of the information that is required to be extracted from the database, they can be made good use of in manufacturing environments. However, the nature of manufacturing is changing towards the need for more flexibility, responsiveness and efficiency from not only the technology but also the manufacturing information system, which has implications for database design. Later development of the relational database methods has offered improvements in this regard.

2.2.3 Relational databases

Hierarchical and network database management systems have performed well over a number of years in manufacturing industry, where data are required to be organized so that the system can respond to highly repetitive requests for information. In effect, this implies that the company can predict with reasonable accuracy the relationships between entities that are required to sustain the business. Thus, the most likely interrogation requirements are determined and the database software written so as to link the relevant relations. Subsequently, the database operator can then navigate through the hierarchy or network with reasonable ease. However, ask a question that requires an unpredicted route through the database and problems will arise. The development of relational databases has provided great flexibility when manoeuvring through data where, because of its organization, it is possible to use simple interrogation statements through accompanying language forms that allow relevant data to be extracted.

The relational database represents the data within a collection of tables that have a direct correspondence to the concept of a mathematical relation. The model was originally proposed by E.F Codd of the IBM San Jose Research Laboratory in the late 1960s (Codd, 1982). The relational model is very efficient at carrying out transactions, and because of the mathematical basis of the model it has been possible to develop efficient algorithms for query processing. The relational system has been developed commercially, and the systems available include Database 2 from IBM, Oracle from The Oracle Corporation, Dbase III and IV from Ashton Tate for PC applications, and PICK from Pick Systems. The relational database grew from the need for data independence, the characteristic that allows data to be changed without recompilation of programs. The relational system is based on the mathematical theory of relations, and as such relational databases may be designed using predicate calculus or algebra. The theory underpinning relational databases is beyond the scope of this book, but the manner in which the data are held is illustrated in Figure 2.5.

In this example, the overall database contains data on three entities: customer orders named **Order list**, suppliers named **Supplier list** and products named **Design list**. Each table is searched by the key attributes (columns of the table), which are **Customer**, **Supplier** and **Product** respectively. The link between

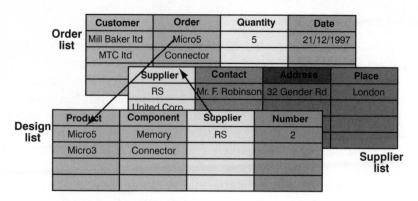

Figure 2.5 *Relational data structure*

the entities, hence the database tables, is established through the same data in the tables: **Order–Product** and **Supplier–Supplier**.

The motivation for developing the relational model was to satisfy three main objectives, found to be lacking in other models, which are:

1. Data independence objective: to provide a clear and precise boundary between the logical and physical aspects of database management.

2. Communicability objective: to make the model structurally simple, so that both users and programmers could understand the database, and therefore be able to communicate easily with one another.

3. Set processing objective: to provide the ability to use a single statement to process multiple sets of records at any one time. Thus, concepts were employed in the relational model that differed significantly from those used in existing systems. One example is *associative addressing*, which allows data to be addressed by value, rather than position. This removes the problems of keeping track of data when positions or locations change, and is also more intuitively appealing to the end user (Codd, 1982).

The relational database uses data dictionaries to store information about the relations, including relation name, attributes, and domains of attributes. The relational system has additional important advantages over other systems. For example, all meaningful relations can be extracted from a relational database, while the other systems only allow extraction where access paths have been previously defined. The relational database is generally inter-rogated using structured query language (SQL).

Being the newcomer to the market, and having been developed from a theoretical foundation, the relational system has had to prove itself in practice in competition with the better-established systems. However, the structure and range of services provided by the database mean that relational systems perform as well as, if not better than, their counterparts.

2.2.4 Object-oriented and hypermedia databases

Conventional DBMS were designed for homogeneous data that can be easily structured into pre-defined data fields and records. Many applications today, however, require databases that can store and retrieve not only structured numbers and characters, but also drawings, images, photographs, voice and other multimedia appli-cations. Manipulating these types of data in a relational system requires extensive programming to translate complex data structure into tables and rows. An object-oriented database stores the data and procedures as objects that can be automatically retrieved and shared. An example of object-oriented hospital database is shown in Figure 2.6.

The hypermedia database approach stores chunks of information in the form of nodes connected by links established by the user (Figure 2.7). The nodes can contain text, graphics, sound, full motion picture or executable computer program. Searching for

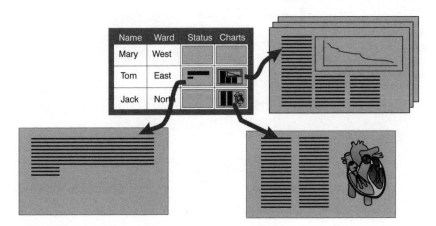

Figure 2.6 *An object-oriented database*

information does not have to follow a predetermined organization scheme. Instead, one can branch instantly to related information in any kind of relationship established by the author. In most systems, each node can be displayed on the screen. The screen also displays the links between the depicted node and other nodes in the database. This database is in essence an object-oriented database with *hypertext* used to search the database and/or to navigate the user through the database. A hypertext is a text that does not contain the information only, but also reference or links to other documents that contain related information. Hypertext is usually highlighted and the operation is activated by clicking on the highlighted portion of the text.

While the object-oriented and hypermedia databases can store more complex data and show better search flexibility, they are slow compared with relational databases.

2.2.5 Physical structure of databases

The way in which data are physically stored in the computer system is different for relational databases and navigational databases. Navigational databases include hierarchical and network databases, so called because the system navigates through the data from one entity to the next.

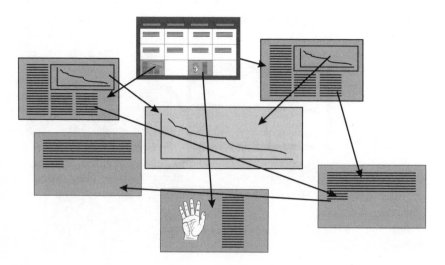

Figure 2.7 *Concept of a hypermedia database*

2.2.5.1 Navigational databases

As we know, a database is a collection of files held on a hard disk. For information to be extracted, the DBMS must be able to find where on the disk the information is located. The search methods for hierarchical and network databases have already been discussed, but would not be complete without looking at exactly what is happening on a practical level.

In addition to the data, each file also contains a number of pointers, each containing the addresses on the disk of other records. In a hierarchical database, each file contains three pointers: to the next record, the prior record and the parent record. In a network database, each file contains this same set of three pointers for each set of which it is a child. This structure is shown in Figure 2.8.

2.2.5.2 Relational databases

Conceptually, a relational database can be seen as a tabular structure of information for which there exists a rigorous method (mathematical set theory) of accessing data. The tabular structure and accessing method will be more thoroughly dealt with in the next section, but the underlying physical structure is fairly simple. The data is held in indexed sequential files, with each file containing a unique identifier, and the files searched until the correct match is found. Although this method is potentially slower than that used in navigational databases, it is made up for by the simpler search methods allowed.

2.3 Database design

2.3.1 Data modelling

The aim of data modelling is to describe a system model that can then be used to create a database containing the same information as the original system. Data modelling theory is as new as the relational model. A *data model* is defined as a collection of mathematically well-defined concepts that help one to consider and express the static and dynamic properties of data-intensive applications (Brodie, 1984). Therefore, a general form for representing a system model would be in terms of entities and their attributes, relationships between entities, and the dependencies of these relationships.

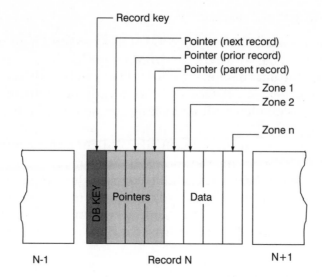

Figure 2.8 *Physical representation of a navigational database*

Data models are essential for the development of information systems, and provide a conceptual basis for analysis. They capture the dynamic and static properties required to support the desired process. *Conceptual modelling* is the process of modelling all properties of an application, and is required to allow the development of application-specific data models. In addition to the static and dynamic aspects, it is also necessary to specify integrity constraints to define allowable states of the database and legal operations. This then allows us to move on to the logical and physical models, as shown in Figure 2.9.

Data modelling produces a schema, defining objects, attributes and relationships, and definitions of transactions. Once these are defined, various attributes can be selected to be presented differently according to the application they are needed for. Figure 2.10 shows the definition of two schemata for an engineer's view and a manager's view of a design database.

To progress from the conceptual model to the logical model, a *data definition language* (DDL) and a *data manipulation language* (DML) are provided for the definition of database schema and the creation of database programs. These functions may be provided within the same language, together with a database query language.

One remaining area of difficulty in database design is the modelling of dynamic operations. It has been recognized that the

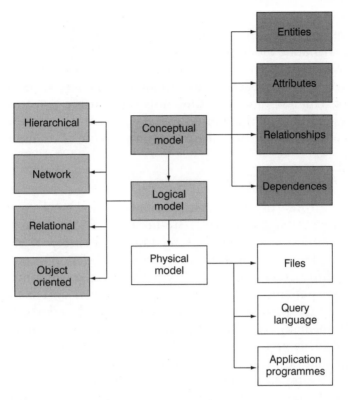

Figure 2.9 *From conceptual to physical data models*

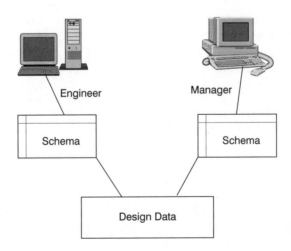

Figure 2.10 *View of data*

provision of such facilities will require significant extensions to existing data models. More recent modelling methodologies include dynamic modelling aspects, such as MERISE, where Petri-nets are used for the dynamic modelling of systems.

A high-level approach to data modelling, i.e., at the conceptual stage of database design, is the entity-relationship construct. This involves defining all the entities in the system and the relationships between them, and is best illustrated through use of examples.

2.3.1.1 Entity-relationship diagrams

There are three possible types of relationships between entities, shown here through examples (Brodie, 1984).

The simplest is a one-to-one relationship between two entities. For example, in a typical hospital there will be a one-to-one (1:1) relationship between a hospital patient and a hospital bed: there will be one patient, and only one sleeping in a bed, and that bed will be, during the patient's stay in hospital, the only bed which she or he will normally use. Diagrammatically, this can be shown as in Figure 2.11. Note the single-headed arrow. The nature of the relationship can be written on the arrow. In the case of the example shown in Figure 2.11, *sleeps-in* could be indicated.

There can be a one-to-many (1:M) relationship between entities. For example, the same patient will be placed in a ward in the hospital, but there will normally be many patients in the same room or ward. In other words, while there is only one ward in which a patient will be placed, there will be many patients within the ward as shown in Figure 2.12. Note the double-and single-headed arrow line indicating that we are expressing many patients in one ward.

Figure 2.11 *Two possible representations of a one-to-one relationship*

Figure 2.12 *A many-to-one relationship*

Finally, there can be a many-to-many (M:M) relationship between entities. A patient–surgeon relationship can be appropriate here. Although this may not normally be the case, a single patient could be operated on by a number of surgeons, especially during a long or complicated illness, and one surgeon could operate on a number of different patients. The diagrammatic relationship is shown in Figure 2.13, with the double-headed format at both ends showing the many-to-many relationship.

2.3.1.2 Entity modelling

A final piece of data modelling terminology that must be known is the term *attribute*. Attributes are the distinguishing characteristics that are held regarding each entity. For the above example of a patient, the attributes held may include aspects such as name, address, date of birth, admittance date, etc.

For a database to operate properly, it is important that each entity has a unique identifying attribute that distinguishes it from all other entities of the same type. Moving back to the example, it may not be sufficient to use patient name as an identifier, as there may be more than one patient with the same name. An attribute such as national health number, where there is no possibility of repetition, may be chosen instead. This unique attribute is known as the *key* to the relation. Once the value of this attribute is known, the value of all the other attributes of the entity are known. This is known as *dependency*. It is also often useful to describe other dependencies, for instance, knowing the name of a stock room and a bin number determines the stock item in a unique fashion.

From such a base description of a system or conceptual model, a logical model can be built up, in accordance with one of the database architectures described earlier. From this stage, the physical model is created, representing the actual database itself.

Attributes can be shown on entity-relationship models as shown in Figure 2.14 where the book entity is taken as an example. Note that the key, very often called *primary key*, of the entity is underlined.

Figure 2.13 *A many-to-many relationship*

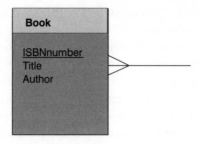

Figure 2.14 *An entity with its attributes*

It is also sometimes useful to indicate whether a relationship is mandatory or optional. An example of such a diagram is shown in Figure 2.15, where an assembly could exist without a supplier (if it is made in-house) whilst a supplier cannot exist (from a database point of view) if they do not supply any product. The relationship is therefore mandatory for the *supplier* entity and optional for the *assembly* entity.

2.3.2 Relational database design

2.3.2.1 General rules

Designing a database is a complex task involving the optimization of a number of different factors in order to produce the best solution. During the systems design, the analyst must balance the trade-off between the following design requirements:

- ease of use;
- speed of response;
- economy of disk space;
- flexibility for future change of the system (Connoly, 1998).

In order to do this, the designer needs to decide on the objectives and priorities of the system, which cannot be done without developing a

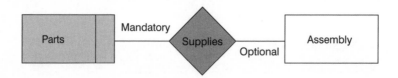

Figure 2.15 *Mandatory and optional relationships*

thorough understanding of the organization for which the system is being installed.

The first stage is to produce a conceptual model of the system as was discussed in Section 2.3.1, deciding which entities need to be included in the system and the relationships existing between them. Following this, the conceptual model needs to be transformed into a logical model, or schema, for the system. In Section 2.2.2, we briefly looked at how this could be done for a network database system, although here we are more concerned with using the relational model.

As described in Section 2.2.3, in a relational database data are stored in tables. Each entry (or occurrence) must be identifiable by a unique key, and the relationships between tables are defined via this key, a key being an attribute that determines all the other attributes in an entity. One-to-one and one-to-many relationships are easy to define in this system. For example, in an ordering system the entities may be **Customer**, **Order** and **Account**, with a 1:1 relationship between **Customer** and **Account**, and a 1:M relationship between **Customer** and **Order**. These types of relationships are based on the logical assumption that a single customer will always have a single account and a single customer can place more than one order at any one time. The tables that would be set up to deal with this situation are shown in Figure 2.16. Note the shaded columns in the tables that are used as key identifiers to provide relationships (links) between tables.

The rules applied here are (Connoly, 1998):

Rule 1: an entity is represented by a relation (table) made up of all that entity's attributes. The identifier is used as the primary key of the table created.

For example, in Figure 2.16, **Name** and **Credit_Status** are attributes of the entity **CUSTOMER**. In this example, the attribute **Name** is a key.

In a similar fashion, **Customer_Name**, **Account_Number** and **Account_Status** are attributes of the entity **Account**, and **Customer_Name**, **Order_Number** and **Value** are attributes of the entity **Order**.

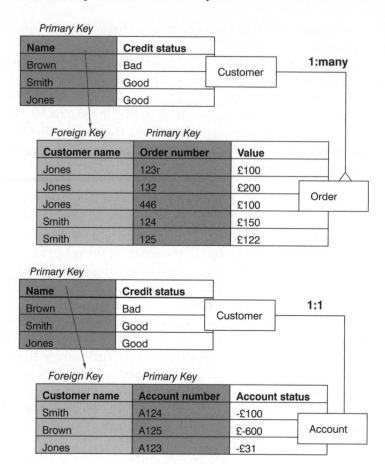

Figure 2.16 *One-to-one and one-to-many relationships: an example*

> **Rule 2: a 1:1 or a 1:many relationship is represented by adding to the relation that represents the second entity (many side) a key of the relation representing the first entity (one side). This key is known as a *foreign* key.**

For example, in Figure 2.16, the relationship (one **Customer**):(many **Orders**) is represented by having the column **Customer_Name** (which is a key of the entity **Customer**) in the **Order** table. Hence, the attribute **Customer_Name** in the entity **Order** is the foreign key. In a similar fashion, the relationship (one **Customer**):(one **Account**) is represented by having the column **Customer_Name** in the **Account** table.

The logical modelling of many-to-many relationships is more complex, and requires the setting up of a third table in addition to those representing the entities. In this way, the relationship can be modelled as two 1:M relationships which, as we have already seen, is a fairly straightforward process. An example of this is shown in Figure 2.17.

The rule applied here is as follows:

Rule 3: in the case of a many:many relationship between two entities, a new relation is created that contains at least one key of each of the original entities, thereby creating two 1:many relationships expressing the many:many relationship. The 1:many relationships are then expressed using Rule 2.

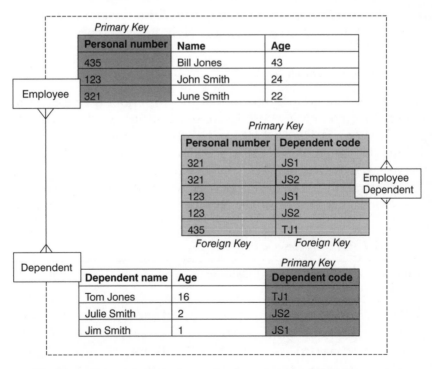

Figure 2.17 *Relational representation of many-to-many relationships*

For example, in Figure 2.17, each employee can have more than one dependant and each dependant can be the dependant of more than one employee. There is, therefore, a (many **Employee**):(many **Dependant**) relationship. A new relation **Employee_Dependant** is therefore created, which contains one of the keys of the **Employee** relation (**Personnel_Number**) and one of the keys of the **Dependant** relation (**Dependant_code**). The (many **Employee**): (many **Dependant**) relationship is now expressed by (one **Employee**): (many **Employee_Dependant**):(one **Dependant**).

In addition to these basics, there are a number of complexities that occur when trying to optimize a system. Consider an attribute that exists for some occurrences but not others. Telephone numbers can be taken as a simple example. Although nowadays most people are expected to have phones, theoretically a system may exist where some people have telephone numbers and others do not. The question is whether to include the **Telephone_Number** as an attribute of the main entity, resulting in a number of empty spaces (hence wasted storage space) on the table, or whether to set up a new table containing all telephone numbers related to the first table. The correct solution does, of course, depend on the specifics of the situation, and this is where the skill and systems familiarity of the analyst comes into play.

2.3.2.2 Reflexive relationships

Sometimes, the conceptual model has to model a single entity linked in different ways. A reflexive relationship or relationship on itself, as shown in Figure 2.19, can be efficiently used to model such a situation. An employee list in a company, as shown in Table 2.5, can be taken as an example.

The hierarchy in the design department is shown in Figure 2.18. It is obvious that the **Employee Number** attribute is the only unique

Table 2.5 *List of employees in a company*

Employee Number	Name	Extension	Department	Job
0009	Tim Jones	2932	Design	Design Manager
0019	Sandra Elmett	3681	Design	Project Manager
0020	Hamill Orad	2210	Design	Design Engineer
0021	Clarrie Steak	2881	Design	Secretary

attribute and hence the primary key. It is also evident that this database can be represented by a single entity, as shown in Figure 2.19, with the reflexive 1:many relationship as every employee can normally have only one superior (boss). If it is essential that this database is capable of showing the hierarchy in the company, the additional attribute **Reports to (Superior)** should be added.

As with any other relationship, the reflexive relationship can be 1:1, 1:m or m:m, and the design rules specified in Section 2.3.2.1 fully apply.

2.3.3 Query languages

A query language acts as an interface between the user and the database. Through it, the user specifies the information required, which is communicated to the DBMS. The DBMS then examines

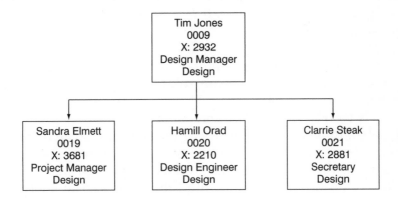

Figure 2.18 *Hierarchy in a company*

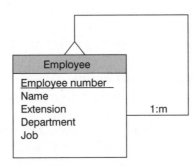

Figure 2.19 *Reflexive relationship*

the database to locate the physical position (on the hard disk) of information required. Throughout this process, the user does not need to know anything about the underlying structure of the database (be it hierarchical, network or relational) or where the information is physically: the user only has to state what is required, and not how to do it.

Although there are a number of different query languages (e.g., DL/1, NDL) the most important to be aware of is *structured query language* (SQL). SQL is a set of English-like commands for defining and manipulating information stored in a database. The first relational database language standard was based on IBM's SQL, which has since been modified and is now an international standard (ISO 9075). Despite standardization, however, there still exist variations between different manufacturers' versions of SQL, and the examples given here are based on the Oracle database and will not necessarily apply to every system. The set of SQL instructions and statements for Oracle Database is given in Appendix 1.

Before data can be accessed, the database itself needs to be defined and data inputted. Although there is no standard SQL command for creating a database, dBase IV SQL and Informix use the following command:

CREATE DATABASE components

where **components** is the name of the database created. As has already been stated, the building blocks of a relational database system are the tables in which the data are stored. Assuming a company divides components into those produced in-house and those purchased from suppliers, a table is required to hold each of these. A table is created by simply using the command:

CREATE TABLE supplied

Now, each column in the table needs to be defined as to the type of data that it will contain. To create a primary key to the table referring to each component by its part number, the command used would be:

partno char(5) not null unique

So, the column is named and all the entries are defined as being five characters long: it is not possible to enter a zero value (every part must have a number) and the final parameter, **unique**, ensures that no two components are given the same number. The remaining columns, for example, part name, supplier and price are then also defined. Here, the supplier is referred to by number to allow easy cross-referencing with the table containing all supplier data. The complete command would be:

CREATE TABLE supplied
(partno char (5) not null unique,
partname char (20),
suppno (5) not null,
price money not null);

Note that the **MONEY** datatype is supported by most implementations, as well as a number of other numerical types such as **INTEGER**, **FLOAT** and **REAL** (see Appendix 1).

Having created a table, it can be destroyed using the command:

DROP TABLE supplied

This should be used with great care, as the deleted table normally is not recoverable.

After the tables have been created and the data inputted, the database is ready for use. The most common command word used in SQL for retrieving data is **SELECT**. This cannot be used on its own, but must be used in conjunction with **FROM**, which tells the DBMS which table to retrieve the data from. For example, the command:

SELECT partno, suppno
FROM supplied;

would give a list of all part numbers and their associated supplier numbers stored in the table **supplied**. As queries often need to be more specific, the **WHERE** operator is also usually required, which can also be used with logical operators **AND**, **OR** and **NOT**. For example:

SELECT partname
FROM supplied
WHERE price < 20 AND suppno = '00247';

would give a list of all parts supplied by supplier 00247 which cost less than £20.

Although it would be impossible to give a complete guide to SQL here (textbooks on the subject run to over 500 pages), the above examples should demonstrate the basics.

2.4 Database normalization

In the methodology presented in Section 2.3.2, we used top-down approach that begins by identifying main entities and relationships. Once a database has been designed, the application of normalization techniques is intended to optimize its performance and storage needs.

The process of normalization is seen as a procedure or series of tests to determine whether it satisfies or violates the requirements of a given normal form (Codd, 1982). Four most commonly-used methodologies for database normalization will be presented in this section: first normal form (1NF), second normal form (2NF), third normal form (3NF) and Boyce-Codd normal form (BCNF). The fourth normal form (4FN) and fifth normal form (5FN) are not very commonly used in practice, and will not be presented here. If interested, the reader is advised to refer to the relevant literature given at the end of this chapter.

2.4.1 The purpose of normalization

Normalization is defined as a technique for producing a set of relations with desirable properties, given the data requirements of an enterprise. Consequently, the process of normalization is a formal method that identifies relations based on their primary key (or candidate keys in the case of BCNF) and the functional dependencies among their attributes. Normalization supports database designers by presenting a series of tests that can be applied to individual relations so that a relational schema can be normalized to a specific form to prevent possible occurrence of update anomalies.

The purpose of database normalization is:

- to minimize duplication (redundancy) of data;
- to eliminate data anomalies that result from duplication.

In short, normalization is a process for assigning attributes to entities.

2.4.2 Data redundancy

A major aim of relational database design is to group attributes so as to minimize data redundancy and thereby reduce the file storage space required by the implemented base relations. The problems associated with data redundancy are illustrated in the following example, which represents three entities of a company database:

Staff_Branch (Staff_Number, Staff_Name, Staff_Address,
Position, Salary,
Branch_Number, Branch_Address, Telephone_Number)

or

**Staff (Staff_Number, Staff_Name, Staff_Address, Position, Salary,
Branch_Number)
Branch (Branch_Number, Branch_Address,
Telephone_Number)**

Here, the names of the entities and concomitant attributes are self-explanatory. In the **Staff_Branch** relation, there is redundant data; the details of a branch are repeated for every number of staff located at the branch. In contrast, if we split the **Staff_Branch** relation into **Staff** and **Branch** relations, the branch details appear only once for each branch in the **Branch** relation.

2.4.3 Data anomalies

There are several anomalies that are to be eliminated by the normalization process:

- insertion;
- deletion;
- modification.

All of these are somehow related to the data redundancy. With redundant data, inserting, deleting or modifying existing data requires this process to be carried out across the whole database.

2.4.4 Functional dependencies

Functional dependency describes the relationship between attributes of an entity (relation), and is defined as follows:

An attribute A1 is said to be functionally dependent on another attribute A2 if knowing A2 defines A1 in a unique fashion, or if each value of A1 is associated with exactly one value of A2.

Taking the example of the **Part** entity, which has the following form:

Part_Number	Part_Name	Supplier
P01823	sheet steel	S678
P98651	plastic cap	S678

we see that the **Part_Name** attribute is uniquely dependent on **Part_Number**. This can be written using established symbology as:

Part_Number → Part_Name

Note that the attribute **Supplier** is not functionally dependent on any of the previous attributes because different suppliers can supply the same part.

Another example is the **Sales_Order** entity of the following form:

Sales_Order_Number	Item_Ordered	Quantity	Customer	Warehouse
PO0001	blue pen	650	Smith	W1
PO0001	red pen	45	Smith	W1
PO0002	blue pen	100	Jones	W1
PO0003	pen set	50	Jones	W2

It is evident that the attribute **Quantity** depends on both **Sales_Order_Number** and **Item_Ordered** attributes. Additionally, **Customer** depends only on **Sales_Order_Number** and **Warehouse** depends only on the **Item_Ordered** attribute. This dependency can be graphically presented as shown in Figure 2.20.

Using the initially suggested symbology, the functional dependencies are:

> **Sales_Order_Number, Item_Ordered → Quantity**
> **Sales_order_Number → Customer**
> **Item_Ordered → Warehouse**

Transitive dependencies are defined as:

If A → B and B → C, then A → C. C is said to be transitively dependent on A, provided A does not depend on B and A does not depend on C.

An example of transitive dependence is the supplier address entity of the following form:

Supplier	Post_Code	Town
PO1823	UB8 1PP	Uxbridge
P98651	SL4 9BQ	Windsor

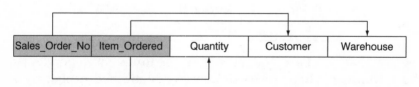

Figure 2.20 *Graphical representation of functional dependencies*

where **Supplier** → **Post_Code** and **Post_Code** → **Town**. Consequently, **Town** is transitively dependent on **Supplier**.

2.4.5 Normalization procedure

2.4.5.1 The first normal form (1NF)

A relation is in the first normal form (1NF) if and only if every attribute is functionally dependent on the primary key. Any table (relation) with a primary key is a 1NF relation.

Taking the entity **Department** of the following form:

Dept_Number	Dept_Name	Location
10	ACCOUNTING	NEW YORK
20	RESEARCH	DALLAS
30	SALES	CHICAGO
40	OPERATIONS	BOSTON

the **Dept_Number** attribute is the primary key in this example because it uniquely determines the other two attributes. Consequently, both remaining attributes, **Dept_Name** and **Location**, are functionally dependent on **Dept_Number**, and this relation is in 1NF. Or, this relation has a primary key (**Dept_Number**), hence it is a 1NF relation.

2.4.5.2 The second normal form (2NF)

A relation is in the second normal form (2NF) if it is in 1NF and no attribute is dependent on only a portion of the primary key.

In the example **Customer_Order** describing sales orders, as

Sales_Order_Number	Item_Ordered	Quantity	Customer	Warehouse
PO0001	blue pen	650	Smith	W1
PO0001	red pen	45	Smith	W1
PO0002	blue pen	100	Jones	W1
PO0003	pen set	50	Jones	W2

with two primary keys **Sales_Order_Number** and **Item_Ordered**, the attribute **Customer** is dependent only on the attribute **Sales_Order_Number** and the attribute **Warehouse** is dependent only on the attribute **Item_Ordered**, that is:

Sales_Order_Number → Customer
Item_Ordered → Warehouse

Consequently, the above relation is not in 2NF, but it is in 1NF (it has a primary key **Sales_Order_Number + Item_Ordered**).

A 1NF table, where the primary key is composed of one attribute only, is 2NF relation by definition. To convert a relation (table) to 2NF, it is necessary to:

■ create a new table for each component of the key and its dependent attributes;
■ create a table for the remaining attributes.

As a consequence, the above example

Customer_Order (Sales_Order_Number, Item_Ordered, Quantity, Customer, Warehouse)

can be decomposed into three new entities, each being in 2NF, as:

Delivery (Sales_Order_Number, Customer)
Packing_List (Item_ordered, Warehouse)
C_Order (Sales_Order_Number, Item_Ordered, Quantity)

Or, in the form of tables, the solution is:

Sales_Order_Number	Item_Ordered	Quantity	Customer	Warehouse
PO0001	blue pen	650	Smith	W1
PO0001	red pen	45	Smith	W1
PO0002	blue pen	100	Jones	W1
PO0003	pen set	50	Jones	W2

=

Sales_Order_Number	Customer
PO0001	Smith
PO0002	Jones

+

Item_Ordered	Warehouse
blue pen	W1
red pen	W1
pen set	W2

+

Sales_Order_Number	Item_Ordered	Quantity
PO0001	blue pen	650
PO0001	red pen	45
PO0002	blue pen	100
PO0003	pen set	50

Note here that in the third table we still have a primary key composed of two attributes, **Sales_Order_Number** and **Item_Ordered**, but the remaining attribute **Quantity** depends on both, not only on one portion. Consequently, this table is, along with the previous two, in 2NF.

2.4.5.3 The third normal form (3NF)

A table is in the third normal form (3NF) if it is in 2NF, and if it does not contain transitive dependency.
   The above entity **Supplier_Address** of the form:

Supplier	Post_Code	Town
PO1823	UB8 1PP	Uxbridge
P98651	SL4 9BQ	Windsor

with the **Supplier** attribute being the primary key, is obviously not in 3NF because it has a transitive dependency: **Town** is transitively dependent on **Supplier**.
   A relation (table) can be converted into 3NF using the following procedure:

- break down transitive dependency;
- store them in separate tables (relations).

The above example of **Supplier_Address** can be converted to 3NF as:
   Supplier_Address (Supplier, Post_Code, Town) =
   Supp_Post_Code (Supplier, Post_Code) + Supp_Town (Post_Code, Town)
Both new tables have a primary key, and no transitive dependency, hence they are in 1NF, 2NF and 3NF. This is exactly what one can easily find in practice: many companies today have a separate database **Supp_Town** with all the postcodes in the UK listed.

2.4.5.4 The Boyce-Codd normal form (BCNF)

To define the Boyce-Codd normal form (BCNF), it is necessary to define the determinant first.

Any attribute or set of attributes whose value determines other values within a row is called a determinant.

Now, the Boyce-Codd normal form is defined as follows:

A table is in BNCF if every determinant in a table is a candidate key. If a table contains only one candidate key, then, if it is a 3NF table, it is automatically a BCNF table as well.

Taking an interview database **Client_Interview** of the following form (Connoly, 1998):

Client_Interview (Client_Number, Interview_Date, Interview_Time, Staff_Number, Room_Number)

Here, the members of staff involved in interviewing clients are allocated to a specific room on the day of interview. However, a room may be allocated to several members of staff as required through a working day. A client is only interviewed once on a given date, but may be requested to attend further interviews at later dates. This relation **Client_Interview** has three composite candidate keys: (**Client_Number, Interview_Date**), (**Staff_Number, Interview_Date, Interview_Time**), and (**Room_Number, Interview_Date, Interview_Time**). These candidate keys overlap by sharing the common attribute **Interview_Date**. We can also select the composite (**Client_Number, Interview_Date**) to act as the primary key for this relation.

Now, the following functional dependencies can be identified from the **Client_Interview** relation:

1. **Client_Number, Interview_Date → Interview_Time, Staff_Number, Room_Number** (primary key).
2. **Staff_Number, Interview_Date, Interview_Time → Client_Number** (candidate key).
3. **Room_Number, Interview_Date, Interview_Time → Staff_Number, Client_Number** (candidate key).
4. **Staff_Number, Interview_Date → Room_Number**.

As functional dependencies 1, 2 and 3 are all candidate keys, none of these dependencies will cause problems for this relation. Also, as there are no partial or transitive dependencies on the primary key (**Client_Number, Interview_Date**), the **Client_Interview** relation is in 3NF. However, this relation is not in BNCF due to the presence of the (**Staff_Number Interview_Date**) determinant, which is not a candidate key for the relation.

2.5 General database management issues

There are a number of database management issues that cannot be taken care of by the DBMS itself. They are, however, very important to organizations utilizing database systems.

The first issue is that of data accuracy. The adage *garbage in, garbage out* applies well to the use of databases, so great care must be taken to ensure that all original data input is accurate, and that any changes are updated on the database immediately.

Security is another important concern. Data may be sensitive, and the economic consequences of a massive loss of information, if the database were hacked into and wiped, could be enormous. It is therefore essential that adequate security measures (passwords, etc.) are set up when the system is initially installed.

An issue that has come to light over the past few years is the social impact of database systems. The fact that a large amount of information relating to people's lives can be stored and accessed is not an idea that everybody is comfortable with. Although this is not strictly relevant to manufacturing databases, in an integrated system including employee details the people factor cannot be ignored. Legislation concerning this area includes the Data Protection Act (1984) in the UK, entitling any person access to data pertaining to themselves held on an electronic system.

The final problem regarding the use of databases to store large amounts of information is that everything held on it becomes encapsulated knowledge. This means that when knowledge is entered into a database system it becomes static; it cannot evolve as further learning is generated unless the database is constantly updated, which is a very time-consuming process. Comparing this to traditional systems, where product knowledge is held by personnel who are constantly learning ways of improving processes, it can be seen that if a database is relied on too heavily there is a danger of adhering to outdated practices, simply because that is what is held on the system.

2.6 Applications of relational databases and future trends

Aside from standard data processing, the relational database has found varied application. For example, it has provided support for graphics and CAD, and much interest has been shown in interfacing CAD packages with database systems. The relational database has also been used in other database applications, for example multimedia databases, knowledge bases, and environment modelling (Korth and Silberschatz, 1991).

For knowledge bases, the aim is to represent facts and rules expressed in logic, which can then be used to answer queries that cannot be handled by basic query languages. The rule processor can interact with the database, thus combining the efficiency of a DBMS and the extended query resolution properties of a knowledge-based system.

One important application of databases is the interfacing with MRP systems. The purpose here is to be able to provide the MRP system with a bill of materials for each component that is scheduled to be produced. Interfacing with a CAD database has historically not been possible as the CAD database does not hold information in a meaningful way, i.e., simply as lines and co-ordinates, whereas a fully integrated system would, at least theoretically, be able to convert incoming orders for products into outgoing orders for the materials and components necessary for those products.

There have been encouraging signs in the development towards integrated databases. Relational databases have a lot to offer to design applications. The use of a relational DBMS for managing graphical data provides several advantages to the applications. Sharing of data between different applications is facilitated. New applications, which must be integrated with existing applications, are easier to develop. It is no longer necessary for graphics application programs to include complex data structures and hence they can be developed by less experienced programmers. Finally, the query language of the DBMS can be used to extract data from the database and use it to draw pictures.

Nowadays, attempts to integrate MRP and CAD systems have been met with some success, although there are still many problems in this area. The first problem is that of data exchange. Although there have been moves towards setting up data standards, such as electronic data interchange and data exchange, for example, there still exists the problem in many organizations of several independent systems run by different functions, without a common data standard. This problem is further compounded by the fact that different functions require the data in different forms; a CAD system that holds a drawing of a component as a series of lines and co-ordinates is of little use to an MRP system attempting to create a bill of materials. Part of this problem could be solved by the introduction of multimedia databases in manufacturing applications. This would allow the current limitation of manufacturing information and data systems based around traditional databases where all data has to be stored in alphanumerical format. A number of prototypes have been developed by research laboratories but large DBMS systems manufacturers have yet to implement the concepts in their industrial applications. Currently, the only applications akin to databases that support multimedia information and documents are groupware systems such as Lotus Notes.

2.6.1 From database concepts to manufacturing applications

As stated in Chapter 1, information is recognized as an important manufacturing resource and a vital factor in successful implementation of new manufacturing technology. Information is today's key resource, and without accurate information, even the most sophisticated plants can sometimes provide a greater quantity of the product least required, at the wrong time. Once the importance of information has been acknowledged, the ensuing process of analysing and reorganizing existing information can bring benefits itself. It is generally accepted that the reorganization of procedures and the redefinition of data paths yield immediate benefits and open up future opportunities for any company. It has been shown in Chapter 1 how this recognition has led to the definition and development of the manufacturing information system (MIS), which attempts to model the manufacturing system in terms of product, process, and information flows. The MIS supports manufacturing functions from shop-floor operations to inventory control, keeps information up to date and allows communication between all levels of the system.

It was also argued in Chapter 1 that MIDS generally addresses three areas:

- planning, including scheduling, materials requirements planning and capacity requirements planning;
- operations control, including shop-floor control and inventory and order management;
- design, including product and plant design.

Designing an information system requires a combination of a top-down and bottom-up approach, to fulfil the objective of the information system.

2.6.1.1 The manufacturing database

A number of authors today stress the necessity for a separate manufacturing database (Lang-Lendorff and Unterburg, 1989). Although a central database for both technical and commercial applications would be the cleanest solution, such a database would be too big and too slow. The manufacturing database mainly contains the parts lists and pointers to drawings and geometric models. The parts data can then be used within manufacturing control systems through software modification routines, which translate the drawing data into manufacturing data. Some authors stress that CIM requires a digitalization of the entire information flow and the creation of a commonly-used data bank. Figure 2.21 illustrates the conceptualization of the integrated information system, covering operations management, design and manufacturing engineering, handling and storage, and manufacturing itself (Major and Grottke, 1987). It is the essence, however, for manufacturing information to be available during the design process, and thus change the product development process from its present rigid sequential structure to a more integrated and parallel activity.

Without a centralized manufacturing database, the number of interfaces required between individual databases grows geometrically with the number of databases. Thus, a single database reduces interfacing problems dramatically, removes problems of data redundancy and inconsistency, and achieves meaningful integration across functional boundaries.

Having one single database is, however, not always technically feasible, economical or justifiable from an organizational point of view, and other structures such as data sharing, organized by central control systems (Figure 2.22), can be used (Waldner, 1992).

Distributed databases are not simply distributed implementations of centralized systems. They provide increased functionality and potentially better performance (see Figure 2.23). A distributed database can be viewed as a collection of data that is logically integrated, but physically distributed across the nodes of a computer network. To this basic definition must be added the notions of *global* and *local* applications. A local application is one in which all the data required are generated, stored, and processed locally. A global application, on the other hand,

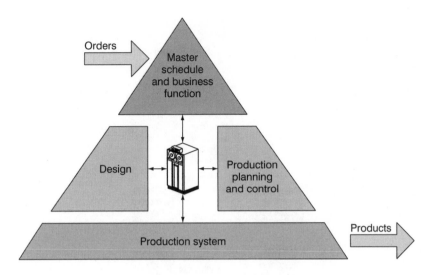

Figure 2.21 *A data structure based on a central source*

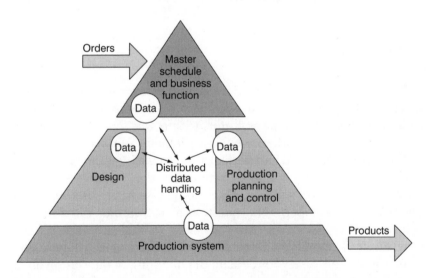

Figure 2.22 *A data-sharing structure organized by a central control system*

is one that requires data that are stored at more than one node. An important characteristic of distributed database management systems (DDBMS) is that each node in the system performs local applications and, in addition, participates in at least one global application. The data at the local nodes are managed by a local DBMS or, in some cases, file management system.

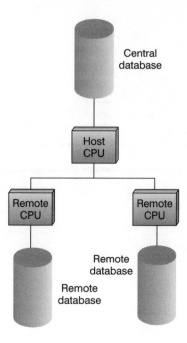

Figure 2.23 *A distributed database*

The advent of improved data communications and increased access to microcomputers has led to a proliferation and diversification of computer systems found within the manufacturing and office environment, with consequent problems of data integrity and redundancy. There are advantages to providing communications between these systems, and to rationalize the available systems to ensure integrity and security. Thus, some system of communication needs to be developed. The question as to whether distributed or centralized databases are most efficient in this situation remains unresolved. To develop a more structured distributed database environment, it is necessary to regulate the handling of queries involving more than one database, and solve general problems of data integrity, such as replication (the process of updating copies, or replicas, of a database across a system of network) and updating of information.

2.7 Conclusion

In this chapter, we have covered the basic structure of a database system and looked at how database management systems have evolved since their inception. The important criteria for choosing a database are:

- speed of access;
- fitness for purpose;
- ease of extracting data.

The basic structure of the three main types of database, hierarchical, network and relational, have also been covered and how the relational model is becoming dominant within industry has also been discussed.

The design of a database is related to the data modelling, ensuring that the database reflects the reality for which it has been designed. Three rules have been defined to convert the concomitant conceptual model into a database physical model in the relational sense. Finally, the database optimization process, known as database normalization, has been discussed.

The relational database is a central part of today's manufacturing information systems. The centralized database has its advantages in reducing the number of interfaces, whereas the distributed database is more optimal in operation.

2.8 References

Brodie M.L (1984). *On Conceptual Modelling: Perspectives from Artificial Intelligence, Databases, and Programming Languages*, Springer, New York.

Burch J.G, Grudnitski G (1989). *Information Systems: Theory and Practice*, John Wiley & Sons, Chichester.

Codd E.F (1982). Relational databases: a practical foundation for productivity, *Communications of the ACM*, 25 (2).

Connoly T (1998). *Database Systems: a Practical Approach to Design, Implementation and Management*, Addison Wesley, Harlow.

Korth H, Silberschatz A (1991). *Database System Concepts*, McGraw-Hill, London.

Lang-Lendorff G, Unterburg J (1989). Changes in understanding of CAD/CAM: a database oriented approach, *Computer Aided Design*, 21 (5).

Major F, Grottke W (1987). Knowledge engineering with integrated process planning systems, *Robotics in CIM*, 3 (2).

Oxborrow E (1990). *Databases and Database Systems: Concepts and Issues*, Chartwell-Bratt, Bromley.

Rothwell D.M (1993). *Databases: an Introduction*, McGraw-Hill, London.

Waldner J.B (1992). *CIM: Principle of Computer-integrated Manufacture*, John Wiley & Sons, Chichester.

2.9 Further reading

Atzeni P (1999). *Database Systems: Concepts, Languages and Architecture*, McGraw-Hill, Cambridge.

Chen P (1986). The entity-relationship model: towards a unified view of data, *ACM Transactions on Database Systems*, 7 (1).

Colton J.S (1988). The design and implementation of a relational materials property database, *Engineering with Computers*, 4 (4).

Deen S.M (1985). *Principles and Practice of Database Systems*, Macmillan Education, Basingstoke.

Occardi V (1992). *Relational Databases: Theory and Practice*, Blackwell Science, Oxford.

Sartori L.G (1998). *Manufacturing Information Systems*, Addison Wesley, Harlow.

3

Manufacturing Resource Planning Systems

3.1 Introduction

The origins of manufacturing resource planning, also known as MRP II, can be traced back to the early 1960s, to a system known as material requirements planning (MRP or MRP I). MRP is an inventory planning and control system, originally running on main-frame computers. It was designed to replace previous techniques for managing inventories that were based on *reorder points* (the minimum stock level at which new stock would be ordered) and *economic order quantities* (the most economic amount of stock to order). The problem with these systems is that they were backward looking, assuming that future demand for an inventory would mirror the past, and could result in an excessive inventory of obsolete stock. In contrast, the theme of MRP is getting the right quantities of supplies and components to the right place at the right time. This is accomplished through the maintenance of accurate databases containing the current inventory and bills of materials (components required for each product), and through the input of the products that need to be made (the master production schedule), the additional materials required can be determined.

During the 1980s and 1990s, the concepts of MRP were extended to cover additional areas of the business. The enlarged version, MRP II, enables systems to examine the engineering and financial implications of future demands on business, as well as the material requirements implications. Hence, the MRP can be described as a total game plan for business (Slack, 2001). Many software suppliers are now looking at integrating wider business functions into *enterprise requirements planning* systems, known as ERP.

3.2 Materials requirements planning

3.2.1 Concept of materials requirements planning

MRP, initially a system in itself, now generally forms the core module of an MRP II system. The required inputs and outputs of this module are shown in Figure 3.1. The algorithm behind MRP is very simple: the demand for parts and components used in manufactured products depends on the quantities in which those products are produced. This concept has been called the *principle of dependent demand*. An MRP system uses the information that relates a manufactured product to its components, for example, how many of each component are required for the product, how long it takes to make each one, and current information on the status of the components, such as how much inventory is on hand, and how many units are in process, to project the number of each component required to support planned future production of the product.

The information aspects of an MRP system will be illustrated through the example of production of a toy car. The relationships and quantities of the parts and sub-assemblies required to make the toy car, commonly called the *bill of materials*, are shown in Figure 3.2

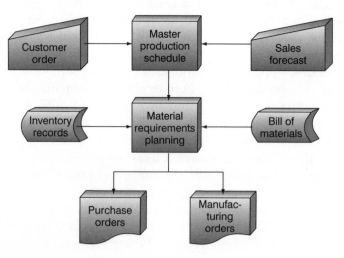

Figure 3.1 *Inputs and outputs of MRP*

(Scott, 1994). The number in brackets in Figure 3.2 shows the quantity needed for parent assembly or sub-assembly.

Each car requires one body-axis mount assembly and two wheel-axle sub-assemblies. At the next level, each wheel-axle sub-assembly requires one axle and two wheels. Therefore, a plan to produce one finished toy car in the future implies that four wheels need to be manufactured in time to assemble them to two axles, which need to be ready for the planned production of the complete car. If it takes two weeks to manufacture wheels and one week to make the wheel-axle sub-assembly, the manufacture of wheels must be started three weeks before the finished car is scheduled for final assembly.

3.2.2 Master production schedule

In information systems terms, the *master production schedule* is a database showing the future requirements for the product quantities and timings. This schedule drives the whole operation in terms of what is assembled, what is manufactured and what is bought. As can be seen from Figure 3.1, it is composed of two components: customer orders, normally referred to as *firm orders* in an MRP system, and forecast demand. For the purpose of this section, the production schedule will be considered to be static and external to the system. In a full MRP II system, the master production schedule is an integrated module. The master production schedule for the toy car example may look as shown in Table 3.1.

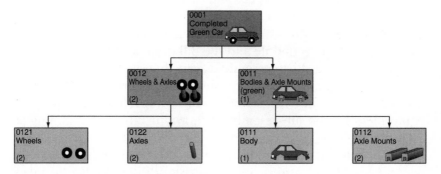

Figure 3.2 *Bill of materials for a toy car*

Table 3.1 *Master production schedule for toy car*

Week	1	2	3	4	5	6
Master schedule, toy cars	10	20	5	0	15	10

From the master schedule for the finished product, production is carried (exploded) down through the product structure to the fabricated or purchased parts level, in order to determine the requirements for these. The information that each car requires two wheel-axis sub-assemblies is held in the bill of materials for that product.

3.2.3 The bill of materials

The data showing which components are required to manufacture a product is known as the *bill of materials* (BOM) for the product. The information contained is identical to that shown in Figure 3.2, the product structure of the toy car, but is held in a different form. The two ways in which MRP systems hold this information are:

- single-level bills of materials;
- indented (multi-level) bills of materials.

In a single-level bill of materials, the details of the relationships between parts and sub-assemblies are held one level at a time, each level showing only the parts that go directly into it. For the toy car, the single-level BOM would be as shown in Table 3.2.

Whilst the single-level BOM is the most common way in which this information is stored, most MRP systems also have the capability to show several levels of the assembly at once in an indented bill of materials. The term *indentation* refers to the indentation of the level of assembly, shown in the left-hand column. The indented bill of materials for the toy car is shown in Table 3.3.

From the master schedule and the bill of materials, requirements for components and sub-assemblies can then be calculated. Using the toy car example, as each finished car requires two wheel-axle sub-assemblies, from the schedule for completed cars the gross requirement for the sub-assembly is shown in Table 3.4 assuming an assembly time is 0.

To translate the gross requirements into net requirements, the inventory on hand and/or orders outstanding need to be taken into account.

Table 3.2 *Single level BOM for toy car*

Part number: 00289

Description: Toy car

Level: 0

Level	Part Number	Description	Quantity
1	10089	Body sub-assembly	1
1	10077	Wheel-axle sub-assembly	2

Part number: 10089

Description: Body sub-assembly

Level: 1

Level	Part Number	Description	Quantity
2	10033	Body	1
2	10045	Axle mount	2

Part number: 10090

Description: Wheel-axle sub-assembly

Level: 1

Level	Part Number	Description	Quantity
2	10031	Wheel	2
2	10027	Axle	1

3.2.4 Inventory records

Three main databases or data sets within a database are generally kept in MRP systems to help manage inventory. These are:

- the item master data;
- the transaction data;
- the location data.

Table 3.3 *Indented BOM for toy car*

Part number: 00289

Description: Toy Car

Level: 0

Level	Part Number	Description	Quantity
0	00298	Toy Car	1
1	10089	Body sub-assembly	1
2	10033	Body	1
2	10045	Axle mount	2
1	10077	Wheel-axle sub-assembly	2
2	10031	Wheel	2
2	10027	Axle	1

Table 3.4 *Total requirements for toy car*

Week	1	2	3	4	5	6
Master schedule, toy cars	10	20	5	0	15	10
Gross requirements, wheel-axis sub-assembly	20	40	10	0	30	20

3.2.4.1 The item master data

These data contains details of all the parts used by the company. Each part is assigned a unique identifier, usually a part number, which can be numeric or alphanumeric. Complex cross-checking methods are often used to prevent errors, for instance, by transposing two digits. In addition to the part number, all other stable data relating to the part are held here, including part description, standard unit of measure and standard cost. Additionally, lead-time to make or buy the part is located here. The danger here is to treat the lead-time as static, as it may vary according to supply conditions or improved processes within the factory. This can result in inconsistencies between the database and the real situation. Updating part lead-time is usually a tedious job as the data are not readily available from the suppliers. The data are usually based on past experience rather than on actual facts.

3.2.4.2 The transaction data

This keeps a record of receipts into stock, issues from stock, and a running balance, so that the current stock levels of any part are known. Historically, transactions were entered into the database overnight or at periodic intervals, but problems of the information lagging behind reality have meant that modern MRP systems run their inventory in real time. This means an increase in the number of computer terminals required and the number of people trained to use them, but the benefits of real-time processing far outweigh this additional cost (Slack, 2001). More on real-time data collection will be given in Chapter 4 relating to the shop-floor data collection.

3.2.4.3 The location data

Whilst some organizations always keep the same part in the same stores location, companies with a wide and changing range of inventory parts find this inefficient, and use a random location system where parts are located in the nearest available place. This is more efficient on space utilization, and also makes it easier to make sure that stock physically turns over by making a *first-in first-out* principle easier to operate. When the computer generates picking lists, it instructs the store operator to pick the oldest stock by sending them to the longest-standing location for that item.

The accuracy of data in the inventory database is incredibly important for the MRP to function properly. For this reason, many companies perform perpetual physical inventory checking, where the actual status of stores is continually monitored against the computer records. This supersedes previous methods where the inventory was audited on an annual basis, as the number of inaccuracies present towards the end of the year caused major problems.

Returning to the toy car example, some of the required components may already be held in the inventory, allowing the net requirements to be calculated, and orders (either work or purchase, depending on whether the part are made or bought in) to be generated.

If 30 wheel-axle sub-assemblies are already in stock, and the lead-time on this component is one week, then the MRP record for this part will have the form as in Table 3.5.

Table 3.5 *MRP record for toy car wheel axle subassembly*

Week	1	2	3	4	5	6
Gross requirements	20	40	10	0	30	20
Inventory/on order	30	0	0	0	0	0
Net requirements	0	30	10	0	30	20
Planned orders	30	10	0	30	20	

3.2.5 The MRP netting process

3.2.5.1 General principle of MRP netting

Whilst the example in Table 3.5 illustrates the MRP calculation for a single level, most products (including the simple toy car example) are composed of multiple levels. These are dealt with by an iterative process where the system moves down through the levels, generating works and purchase orders for each level and using them to supply the master schedule for the next, as shown in Figure 3.3.

Again, available inventory of each item is checked, and works orders are generated for the net requirements that are made in-house, as are purchase orders for the net requirements of items that are bought from suppliers. Moving down to the next level for the toy car, each wheel-axle sub-assembly requires two wheels, and wheels have a two-week lead-time, etc.

Allowing for the lead-time on parts and calculating when the order needs to be placed is known as *back scheduling*, as in the above examples.

3.2.5.2 MRP netting using MRP charts

A more common way of MRP netting is by using MRP charts. Most of MRP software packages today use this method. The principle of using MRP charts for MRP netting is also an iterative approach based on a level-to-level analysis of the quantity demand and will be explained in an example of manufacturing toy cars.

The bill of materials for a toy car, along with the lead-time for every component and sub-assembly, is given in Figure 3.4. The figures in brackets indicate the quantity needed for the parent assembly.

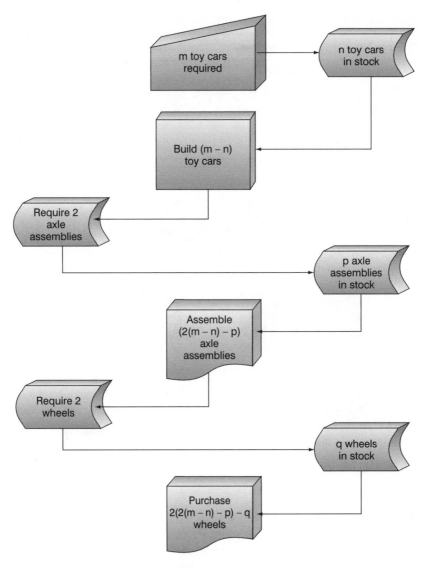

Figure 3.3 *Example of an MRP netting process*

Additionally, the current stock level of components and sub-assemblies, as well as required safety stock is given in Table 3.6. Here, the safety stock is used to compensate for unexpected demand.

In its simplified form, the requirement from the master production schedule is to deliver 40 completed toy cars in week 8.

Table 3.6 *Available stock level and safety stock for toy car*

Part number	Part name	In stock	Safety stock
0001	Completed car	10	5
0011	Bodies and Axle Mounts	10	5
0012	Wheel and Axles	10	10
0111	Body	5	5
0112	Axle mounts	5	10

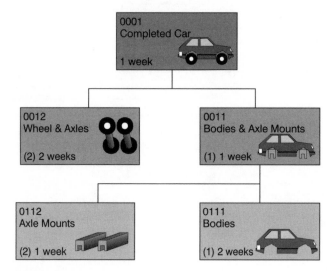

Figure 3.4 *The toy car bill of materials*

The netting process always starts from the highest level of manufacturing, in this case from the complete car component. Hence, it is important to deliver 40 toy cars at week 8. As we have 10 completed cars already on stock, and we have to keep 5 completed cars as safety stock, it is necessary to place the order for 35 cars at week 7 (the lead-time for a complete car is 1 week, see Figure 3.4), as shown in Table 3.7.

The order of the toy cars at week 7 will, in turn, initiate the requirements for the next level components/assemblies: wheel and axes and body and axes mounts. Taking into account lead-time and the quantity required from the bill of materials, as well as the available stock and requested safety stock, the concomitant MRP lists are shown in Table 3.8. Similarly, the requirements for 30 body

Table 3.7 *MRP record for a complete toy car*

0001 – Complete car	1	2	3	4	5	6	7	8	9	10	11
Gross requirement								40			
Scheduled/planned receipts								35			
On hand	10	10	10	10	10	10	10	10	5	5	5
Planned orders							35				

and axle mounts at week 6 will trigger the requirements for the components at the next production level: axle mounts and bodies, as shown in the MRP lists below.

3.2.6 Closed-loop MRP

The MRP system described above is an open-loop system, with no feedback mechanism to say whether the plan is achievable or whether it has actually been achieved. This was how the original MRP systems operated, with materials plans being launched at the start of the week, and a complete replanning exercise taking place the following week launching a new set of plans.

Improvements to this system have involved closing the planning loop, meaning that production plans are checked against available resources. Capacity is checked throughout the process, and if the proposed plans are not achievable at any level they are then revised. The three planning routines used to check production plans against the operations resources are:

- resource requirements plan;
- rough-cut capacity plan;
- capacity requirements plan.

Closed-loop MRP can be refined to a very short-term plan, and all the levels are illustrated in Figure 3.5.

3.2.6.1 Resource requirements plan

These are static level plans that involve looking forward in the long term to predict the requirements for large structural parts of the operation. This would include aspects such as the numbers of location and sizes of new plants. They are sometimes referred to as *infinite capacity plans* as they assume an almost infinite ability to

Table 3.8 *MRP lists for toy car subassemblies and components*

0012 – Wheel and Axles	1	2	3	4	5	6	7	8	9	10	11
Gross requirement							70				
Scheduled/planned receipts							70				
On hand	10	10	10	10	10	10	10	10	10	10	10
Planned orders						70					

0011 – Body and Axle	1	2	3	4	5	6	7	8	9	10	11
Gross requirement							35				
Scheduled/planned receipts							30				
On hand	10	10	10	10	10	10	10	5	5	5	5
Planned orders							30				

0111 – Body	1	2	3	4	5	6	7	8	9	10	11
Gross requirement						30					
Scheduled/planned						30					
On hand	5	5	5	5	5	5	5	5	5	5	5
Planned orders					30						

0112 – Axle Mount	1	2	3	4	5	6	7	8	9	10	11
Gross requirement						60					
Scheduled/planned receipts						65					
On hand	5	5	5	5	5	5	10	10	10	10	10
Planned orders						65					

increase production in line with demand. This is because they are attempts at making the required resources available to facilitate long-term production plans. An example of this is a production plan to manufacture 3000 toy cars a month. The feedback loop would comprise of the questions *can we do this?* and *what resources do we need?*

3.2.6.2 Rough-cut capacity plan

In the short to medium term, only existing capacity is available for production. At this level, the feedback loop checks the master

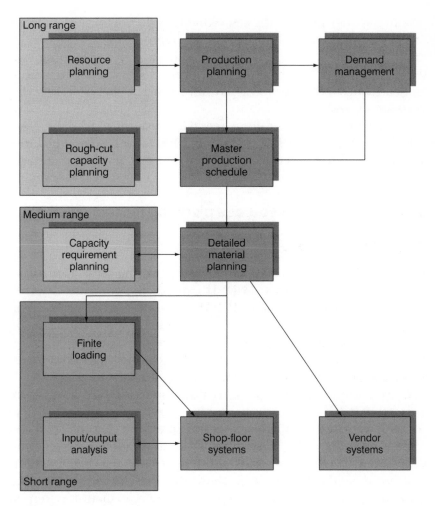

Figure 3.5 *Theory behind closed-loop MRP*

production schedule against known capacity bottlenecks and key resources. These are finite capacity plans because they need to operate within constraints, and if the master production schedule is not achievable then it will need to be adjusted. The feedback question for this level may be *can we make 400 toy cars for week 35?*

3.2.6.3 Capacity requirements plan

The capacity requirements plan projects the actual work orders ahead to predict the day-to-day loading of machines. These are infinite capacity plans in the sense that they do not take the capacity constraints of each machine or work area into account. If the load is lumpy, i.e., uneven, then it may be smoothed by replanning to a finite capacity or by allocating temporary resources to the area. A feedback question at this level may be *can we make 400 wheel-axle sub-assemblies by week 33?*

The addition of closed-loop feedback systems marks the transition from MRP to MRP II . However, a full MRP II system includes a number of other functions, which are detailed in the next section.

3.3 Manufacturing resources planning system (MRP II)

3.3.1 MRP II functions

Whilst MRP is essentially aimed at planning and control of production and inventory, MRP II extends the concepts to other areas of the business. Hence, MRP II can be defined as a game plan for planning and monitoring all the resources of a manufacturing company: manufacturing, marketing, finance and engineering. Technically, it involves using the closed-loop MRP system to generate the financial figures (Wallace, 1990).

The information flows in an MRP II system are shown in Figure 3.6. An MRP II system is not only a software package: it also contains the organizational side of the enterprise. Figure 3.6 shows what is typically done by software systems (shadowed boxes) and what is not. It is important to differentiate between the functions that are indeed computerized, those that are not, and those that could be. This will, of course, depend on the information system in place in the company.

It becomes apparent from the diagram in Figure 3.6 that the master production schedule, previously considered to be externally generated in an MRP system, is an integrated part of MRP II being generated from the interaction with other business functions. One of the problems with early systems was instability or nervousness within the master schedule: changes in demand would cause a shift within the master production schedule, resulting in plans changing faster than they could be executed. This problem was solved with the introduction of a *planning time fence*. This is the time period within which it is very difficult to change the schedule, generally corresponding to the manufacturing or assembly lead-time, and

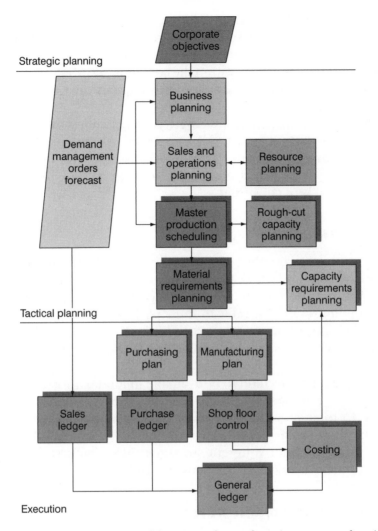

Figure 3.6 *Typical computerized functions of manufacturing resource planning*

can vary for different groups of products. Beyond the time fence, the system plans and rescheduled orders in support of demand and customer orders, whilst within it the system will advise the (human) planner of any problems (in the form of exception messages) and after analysis the planner decides whether to action or ignore.

What also becomes apparent from Figure 3.6 is the manner in which an MRP II system is integrated across a wide range of company functions. It is no longer a production/inventory-centred system, as is the case with MRP, but includes top management (setting corporate objectives), sales and marketing (demand management) and finance. The benefit of this system is that rather than different packages being used by different functions within the company, the whole company is tied together by a single system primarily implemented in the form of a software package.

Resulting from the use of a single package comes an additional and very important result. All the separate company functions utilize a common database. This avoids discrepancies between the use of different databases, and problems of attempting to ensure all databases are kept consistent.

3.3.2 MRP II software

There are a large number of commercial MRP II software packages on the market, and a task for any company implementing an MRP II system is to decide which is most suited to their company. Most software houses do not refer to their packages as MRP II systems, but rather as *manufacturing systems*. This is in part due to MRP/MRP II having received some bad publicity, but the principles operated on are the same.

Software packages tend to be available in modular form, allowing companies to implement all the modules that they feel are appropriate to their organization. Many packages also offer a *what if?* facility, allowing various production plans to be simulated and the results examined for different strategies, along with various techniques for extracting and analysing information.

3.3.3 The benefits of MRP II systems

Companies using MRP II successfully have claimed a wide array of benefits to have resulted from the system. These can be summarized as:

- increased customer service and satisfaction;
- improved utilization of facilities and labour;
- better inventory planning and scheduling;
- faster response to market changes and shifts;
- reduced inventory levels without reduced customer service;
- reduced purchase costs;
- effective engineering changes.

However, not all companies implementing such systems realize all these benefits and evidence. The statistics show that many companies fail. Much promotional, but also governmental literature often inflates the benefits achievable by MRP II, and any system should be justified carefully, as with any other information system.

One way of measuring the success of an MRP II implementation is the *ABCD checklist* developed by The Oliver Wight Organization, who are one of the leading MRP II consultancies (Wallace, 1990). A grade is awarded to the company depending upon the answers supplied to the checklist, which can be broadly summarized as:

Class A: effectively used company wide, generating significant improvements in customer service, productivity, inventory and costs.

Class B: supported by top management, used by middle management to achieve measurable company improvements.

Class C: operating primarily as better methods for ordering materials, contributing to better inventory management.

Class D: information inaccurate and poorly understood by users, providing little help in running the business.

Given that there is such a wide range in received benefits from these systems (from very successful to wasting money), a brief examination of the implementation of such systems is necessary.

3.3.4 Implementing an MRP II system

3.3.4.1 Managing the implementation

A number of studies have been made into the implementation of MRP II and, from examining these, a number of common themes emerge (Roberts and Barrar, 1998).

The first is that MRP II is not a panacea for solving the problems of a business. Benefits do arise from the use of these systems, but only if a large amount of effort is put into ensuring that they operate correctly. Following from this is the need for the project to have been initiated from business requirements, i.e., corporate strategy. This allows the exact organizational requirements to be defined from which operational goals can be set.

MRP II is also a people-based system, not a computer-based one. Companies believing that they are implementing a computer system to automate inventory planning and scheduling get just that, which puts them in category C of the Oliver Wight ABCD classification. Ideally, implementation involves the whole company and changes the way in which the company operates. Education and training programmes are essential for all personnel, and strong involvement from end-users during the project helps establish ownership from them, resulting in greater commitment to ensure that the system succeeds.

All MRP experts argue that MRP II systems also need to be implemented with full support and involvement from top management. This is important so that resources required are made available, and so that personnel appreciate the importance of the system. The existence of a steering committee for the implementation has also been positively linked to success.

Best practice for implementing MRP II has been defined by Oliver Wight Companies and their results, developed and refined in the late 1980s, have been given the name *The Proven Path* (Wallace, 1990). Their system calls for an aggressive implementation, taking a maximum of 18 months to two years, and the steps defined are shown in Figure 3.7.

3.3.4.2 Data accuracy

Successful implementation requires high data integrity in the MRP II database. Inaccurate data leads to poor performance from the system (*garbage in, garbage out* syndrome) and a loss of faith in the results generated, and hence the system itself.

Although the importance of data accuracy is often difficult to explain, there is not much point in trying to use MRP under 95 per

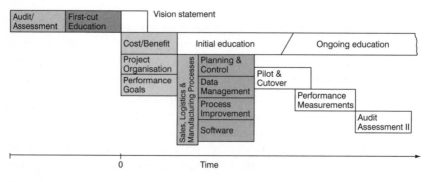

Figure 3.7 *The proven path*

cent data accuracy for all records (Luscombe, 1993). This can easily be explained. Because the bill of materials explosion process is at the centre of MRP, any inaccuracy at one level will translate into inaccuracies at lower levels. Looking at the bill of materials in Figure 3.8, for instance, an inaccuracy of 10 per cent in the stock levels for 'bodies and axle mounts' records will translate into inaccuracies when calculating the requirements and therefore any future stock holdings for 'bodies and axle mounts', making half of the stock items inaccurate by 10 per cent. Some authors recommend that bills of materials should be at least 98 per cent accurate, with the accuracy figure calculated as (Clement, Coldrick and Sari, 1992):

$$Accuracy\ of\ BOM = \frac{number\ of\ misses}{number\ of\ audits} \qquad (3.1)$$

An *audit* is carried out for products in the BOM and scores a *miss* if any of its components or parents are either:

- wrong (quantity per or unit of measure);
- missing or present in error.

For example, an audit of the 'wheels and axles' component of Figure 3.9 would also count as a miss in the cases illustrated in Figure 3.10 and Figure 3.11.

Ideally, the users should be made accountable for the accuracy of data they input, and frequent reviews of data undertaken both to keep inventory records accurate and to review lead-times which, in a company committed to continuous improvement, should be constantly changing.

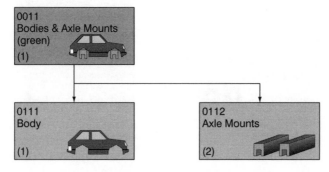

Figure 3.8 *Example of the bill of materials*

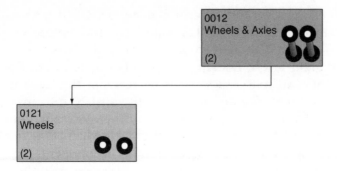

Figure 3.9 *Audit miss (missing child component)*

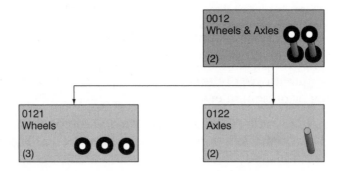

Figure 3.10 *Audit miss (wrong number off in child component)*

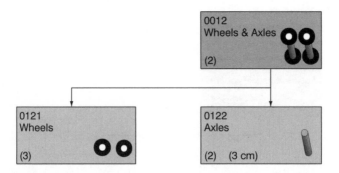

Figure 3.11 *Audit miss (wrong unit in child component)*

3.3.4.3 Designing the bill of materials

Ideally, all bill of materials, issues and inventory should be formalized prior to the implementation process. There are obviously a number of ways that a company can choose to build its bill of materials and choose to code the components. The technique chosen will depend on a number of requirements, such as:

- the frequency of engineering changes;
- routings;
- ease of maintenance;
- proportion of customized products;
- reprocessing.

It is worth being familiar with two terms often used when talking about a bill of materials design: phantom components and modularization.

Phantom components are components that are not actually stocked and which are therefore ignored by the netting process. For instance, in Figure 3.9, assuming that the wheels are clipped on the axles when required for the completed car, there is no point in planning for the 'wheels and axles' component. It could, nonetheless, be useful to keep in the bill of materials for engineering or costing purposes. The component could therefore be defined as a 'phantom', with a lead-time of 0. Most MRP software packages have facilities to *blow through* these components such that no requirements are generated for them.

Modularization is used in cases where there are a large number of options available for the final product. Instead of creating a bill of materials for each possible permutation of options, a bill of materials is created for the common components and these sub-assemblies are then used as phantom sub-assemblies.

3.4 Conclusion

We have seen how MRP systems have evolved from automated methods for process planning and inventory control to company-wide information systems encompassing all functions. The basic logic on which these systems are based is fairly simple, but requires high data integrity and full commitment from the organization if they are going to be successful. There are a number of success levels at which the MRP II can operate, as defined by the Oliver Wight ABCD classification. There is also a wide range of commercially available software that offers MRP II capability. A strategic analysis of the organization's business requirements is necessary to select the system that is best suited to them and fulfils their needs.

3.5 References

Clement J, Coldrick A, Sari J (1992). *Manufacturing Data Structures: Building Foundations for Excellence with Bills of Materials and Process Information*, Oliver Wight Ltd.

Luscombe M (1993). *MRPII: Integrating the Business, a Practical Guide for Managers*, Butterworth-Heinemann, Oxford.

Roberts and Barrar (1998). MRP II implementation: key factors for success, *Computer-Integrated Manufacturing Systems*, 5 (1) pp 31–38.

Scott B (1994). *Manufacturing Planning Systems*, McGraw-Hill, London.

Slack N (2001). *Operations Management*, Pitman Publishing, London.

Wallace T (1990). *MRPII: Making it Happen*, Oliver Wight Ltd.

3.6 Further reading

Attaran M (1992). *OMIS: Operations Management Information Systems*, John Wiley & Sons, New York.

Burns R (1991). Critical success factors in Manufacturing Resource Planning implementation, *International Journal of Operations & Production Management*, 11 (4) pp 5–19.

4

Shop-floor Data Collection Systems

4.1 Introduction

Raw data are at the heart of all information systems. In a manufacturing environment, a large amount of data is generated on the shop floor itself, and shop-floor data collection represents the way this data is to be collected in order to improve shop-floor performance. After examining exactly what is meant by shop-floor data collection and why we would wish to do it, the technologies used in this area will be examined in detail, from bar codes to voice recognition systems, finishing off by looking at how people may react to the introduction of this technology.

4.1.1 Definition of shop-floor data collection

A sample of manufacturing managers' definitions of *shop-floor data collection* (SFDC), collected by the IB Consulting group survey, included replies such as 'collecting data from the factory floor', 'collecting and presenting information on machine status, staff attendance, quality losses, scrap, etc.' and 'networked information system available to management and operators and providing real-time and historical process data' (Quinn, 1994). In essence, SFDC is about finding out exactly what is happening on the factory floor as a starting point for improving manufacturing performance. Shop-floor information includes:

- process yield and scrap;
- machine performance and usage parameters;

- operations time;
- order status;
- inventory and product traceability;
- quality data;
- personnel.

4.1.2 Rationale for shop-floor data collection

There are a number of different reasons for collecting data from the shop floor, with different people requiring different information in order to meet their functional objectives. For example, SFDC may focus on the manufacturing process, in order to improve utilization, throughput and scheduling, or alternatively the focus may be on the product in order to provide traceability and ensure high quality. The actual data required and the time lapse between the generation of the data and their use vary widely, depending not only on who needs them and what they need them for, but also how they have been generated. Table 4.1 illustrates the varying requirements of different users. At one end there is the operator who requires data from his machine within seconds of their generation, whilst at the other is the senior manager who is more concerned with the long-term implications of manufacturing performance on a macroscopic scale.

Table 4.1 also illustrates the wide variance in the different types of data required, ranging from the location and condition of a specific part within the system to utilization of a particular machine.

Finally, it must be remembered that collecting data is a non-value-adding activity; manufacturing exists to create products,

Table 4.1 *SFDC requirements*

Users	Typical Needs	Timeliness
Operator	Machine data	Seconds
Team leader	Work tracing	Minutes
Line manager	Throughput, shift reports	Hours
Engineering	Yield, machine performance	Days
Production planning	Inventory, work tracing	Days/weeks
Finance	Usage	Weeks/months
Senior management	Management data	Months

not data. For the activity to be worthwhile, the data must be acted on to create value by improving the processes. Therefore, collection must ensure data value, data must be shared and data must be used.

4.1.3 Methods of shop-floor data collection

There are a number of different methods by which SFDC can be performed. The simplest, and cheapest, is paper recording and manual storage. This method makes it fairly difficult to use and analyse the data, hence there is a greater probability that the data will not be used to improve the process, making the exercise pointless.

The second method is paper recording and input into an MRP system. Although this is cheap to perform, it is labour intensive, resulting in a time lag, low accuracy and is also difficult to analyse.

The next option is to use shop-floor terminals linked to an MRP system. Although this can only be used for manpower and material tracking and also has a time lag, it has better accuracy than the previous method.

Finally, dedicated shop-floor data collection systems can be implemented that are very flexible, very accurate, and allow the possibility of providing information in real time. These are the computerized systems examined in detail in the next section.

4.2 Computerized SFDC

A *computerized data collector* is defined as an independent entity that captures, stores, processes and forwards data to a host computer, as shown in Figure 4.1 (Cohen, 1994). From this definition, there are four basic features of a computerized data system:

- means of inputting data;
- memory capacity for storing data;
- independent processing capability;
- data communication to a host system.

There are a number of different technologies that can be used to provide these four features, with the choice being dependent on the local constraints upon the system.

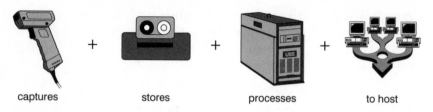

captures   stores   processes   to host

Figure 4.1 *Concept of computerized SFDC*

4.2.1 Technologies for SFDC

The technologies available for fulfilling the four functions listed above are shown in Table 4.2.

The remainder of this chapter concentrates on the differing types of data collection devices, whilst information on distribution, storage and processing technologies is to be found elsewhere in this book.

At the data collection point, there is a need for some form of terminal. This can be either fixed in position or of the portable variety, and there are a wide array of both types available on the market, with the final choice depending on the system requirements, and especially the collection technology used. An example of a complete data collection system that includes portable terminals is shown in Figure 4.2. Notice the variety of interfaces required for the terminal to integrate with various other components of the information system.

As portable terminals require a very light power source, they run low-power processors, and as a result of this they are not particularly fast. The practical effect of this is that the code run needs to be optimized, in order to provide satisfactory operating speeds.

Table 4.2 *SFDC technologies*

Function	Technologies
Collection	Keyboard input, bar code, cards, PLC links, device controllers links
Storage	RAM, bubble memory at source, floppy/hard disk, centralized storage
Processing	At terminals level, PCs, client/server, mainframes
Distribution	RS232, local area networks, radio link

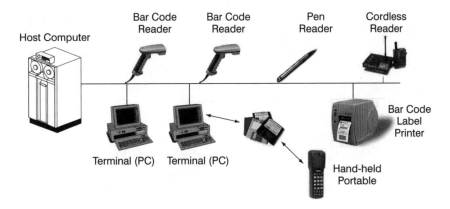

Figure 4.2 *System diagram for shop-floor data collection*

4.3 Bar codes

4.3.1 Introduction

The most popular data collection device is the *bar code reader*, used to decode a bar code in order to produce a binary code. A bar code is a pattern of rectangular bars and spaces of varying width (Figure 4.4). The use of bar codes as a method of data collection has progressed dramatically since their inception at the start of the 1970s, notably with the adoption of the UPC code in the USA in 1973, although it was not until 1982 that the US Defence Department adopted the now industrially widespread *code 39*.

Bar code technology still represents 60 to 70 per cent of applications and continues to be improved with the advancement of other technologies (Pons and Chevalier, 1993). These have become more flexible in order to favour the simple integration of complex systems and to assure compatibility between different identification systems and the interchangeability of products from different manufacturers. Progress in portable code readers allows their operation in real time due to technologies of transmitting computer data without cables by radio or infrared communication. There is also the miniaturization of the data supports and more dense and evolving information systems.

Bar codes can be put to a wide variety of uses in shop-floor data collection systems. At the simplest level, bar codes can be used to automatically identify products as they pass a certain point on the line. More complex systems can also be used to record utilization of equipment and personnel, where bar codes are printed on the work-order forms and employee identity cards. These systems allow data to be collected regarding worker identity, work-order number, task number, start and end times of task and quantities. From this, a much more complete picture about what is happening where and when on the shop floor can be generated.

A further example of the usefulness of bar codes is their use in warehouse management systems with a hand-held cordless bar code reader (Figure 4.3). This provides extremely accurate inventory control, as everything entering or leaving must have its bar code scanned. This can speed up the manufacturing process, as stock and products are easier to retrieve due to the exact quantities being in the precise location expected, and it also allows a reduction in operational inventory. In addition, some systems also allow optimization of storage space by recommending storage locations for particular items.

4.3.2 Bar code technology

There are a number of standard symbologies for data encryption in bar code format, some being fully alphanumeric, such as code 39, others partially alphanumeric containing only some letters, and most of them are numeric only. Figure 4.4 shows some of alphanumeric and numeric-only barcodes.

In order to distinguish between different symbologies, there is some basic terminology that needs to be understood. The first of these terms refers to the thickness of the bars and is known as the *X dimension*. The X dimension is the basic unit of thickness of each of

Figure 4.3 *Cordless hand-held bar code reader*

Figure 4.4 *Several types of barcode: starting from upper left corner – Code 39 (EXAMPLE – alphanumeric), Interleaved (123456789 – numeric), EAN-8 (1234567 – numeric), EAN-13 (12345678910 – numeric), UPC-E (123456 – numeric) and UPC-A (12345678901 – numeric)*

the bars, so every bar in the code will be of thickness X or a whole multiple thereof, for example 2X or 3X. Obviously, the value of the X dimension determines the physical length of the code for a specific piece of data, or the information density that can be carried by the code.

The second characteristic of a symbology is whether it is discrete or continuous. In a discrete symbology, a light bar or space of a predetermined width is left between each character, whilst in a continuous symbology characters always start with a dark bar and end with a light one. Continuous symbologies allow for shorter bar codes, whilst discrete ones provide greater data integrity.

A third characteristic of bar codes is the *quiet zone*; a minimum area of light space that must precede and follow the actual code in order to enable a successful read. It does this by providing the reader with a reference as to what constitutes light and dark. Its length is determined by a multiple of the X dimension and varies between symbologies.

In addition to data, bar codes also include control characters. These include start and stop characters, which enable the code to be read forwards or backwards. If the stop characters are read first, then the following data are reversed during decoding. An example of a basic bar code structure is shown in Figure 4.5.

Some bar codes also include a check digit, which is usually the final character of the code, in order to improve data integrity. The characters in the bar code are passed through an algorithm,

| Leading quiet zone | Start character | Data ('1') | Stop character | Trailing quiet zone |

Figure 4.5 *Bar code structure*

an arithmetic process done by the decoder, and if the result is the same as the check digit then the system knows that it has read the code correctly.

The technology needed to produce bar codes is fairly simple. They can be printed on a variety of printers, including dot matrix, laser, thermal and ink jet, and depending on the information needing encoding, bar codes can either be printed in advance or at the point of application. Dedicated printers for on-site or other printing of bar codes are also commercially available (Figure 4.6). Several specifications for bar codes can be given regarding their appearance, including dimensional (amount of tolerance in width variation), printing (spots, voids and edge roughness) and contrast, with the actual specifications being dependent on the type of reader and expected operating conditions. Contrast is important regarding labels, as some colour combinations are not permissible. Whilst good contrast is provided by combinations such as black on white or green on yellow, yellow on white or black on blue is not sufficient.

Figure 4.6 *Specialized bar code printer*

4.3.3 Bar code readers

Traditional bar code readers operate by reflecting a light source off the code and decoding the reflected pattern in order to capture the data in a digital form. The reading process has a number of stages. The reader emits light, which is reflected by the white bars but absorbed by the dark bars of the code. The reflected pattern is converted into a fluctuating electrical current when it is picked up by a photodiode in the reader. This analogue signal is then converted into a binary digital signal by a circuit called a *wave shaper* (Figure 4.7). The decoder then converts this into the ASCII characters determined by the particular symbology used.

Bar code readers can capture information by contact or at a distance, and can be manual or automatic. The best-known manual readers, and the most widely used by operators, are the laser handheld readers and the light pen (Figure 4.8). There are several models available, high density or low density, visible light or infrared, each having different characteristics of use regarding range, speed of reading, first read rate, substitution error rate and resolution. The resolution, which determines the density of the bar codes that can be read, is dependent on the size of the light spot emitted by the reader.

Longer distance manual readers use a laser diode. These are more expensive then light pens but have a number of advantages, including a much higher first read rate due to the rapidly-moving laser beam performing several scans per second and, with some configurations, the ability to read a code at any angle. The more coherent source also allows denser bar codes to be used, increasing the amount of information that can be carried in a given area.

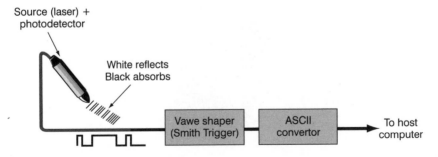

Figure 4.7 *Bar code reading process*

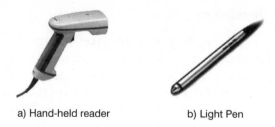

a) Hand-held reader   b) Light Pen

Figure 4.8 *Bar code readers*

Laser bar code readers are available in either hand-held or fixed mountings (Figure 4.9).

Another widely-used method of bar code capture is the *charge coupled device* (CCD) scanner. Broadly speaking, this falls slightly below the laser scanner in range and ability to read obscured/damaged codes, but has great advantages in terms of robustness and power consumption, often making it the scan-engine of choice for portable data collection terminals (Figure 4.10). Robustness is achieved by a wholly solid-state construction, replacing the mechanically-scanned beam of the laser scanner with an imaging system that takes a linear picture of the entire bar code and reads it end to end electronically.

4.3.4 Characteristics of bar codes

The use of bar codes in shop-floor data collection has a number of advantages and disadvantages associated with it. On the positive side, this is a simple technique that is exact and reliable, eliminating the reading or mis-keyed errors that may occur in a manual entry

Figure 4.9 *Fixed laser diode based bar code reader*

Figure 4.10 *CCD bar code reader*

system. There is also an international standard to ensure greater compatibility between systems (e.g., EAN-13 for the grocery trade in Europe), and the technology is both well established and widespread, as well as being extremely versatile and easily upgradeable. The speed of entry data seizure is high so the technology can cope with high flow rates and, finally, the capital and running costs involved are low.

On the negative side, bar code labels are easily damaged in severe environments, for example, in the food processing industry. The information encoded is also fixed and non-evolving, and cannot be updated as the part, for example, has further processes carried out on it; instead a new label must be printed. A further disadvantage is that very often an operator is required for the scanning process. Bar codes also have a very limited data capacity; a typical bar code never extends to more than 20 characters and is usually shorter, in order to keep the code to a reasonable physical length. Bar codes are not recommended as a way of encoding sensitive data due to the widespread availability of bar code readers. Finally, visibility is required without any intermediary obstacles between the bar code reader and bar code itself, and the reading distance is somewhat limited, often to several metres.

4.4 Electronic labels

Also known as *radio frequency* (RF) *tags/transponders*, these are systems where information is transferred between the identified objects containing the label or electronic chip, and the processing system via radio wave changes in the electromagnetic fields.

The principal application of this technology is the identification of people and objects, for example, for controlling access to areas or following the progress of production materials.

4.4.1 Implementation

The two main components of an electronic labelling system are the tag itself and the interrogator that communicates with the tag and acts as the interface to the host system (Figure 4.11).

The tag is a transponder that contains one or more chips, a receiving antenna and a transmitting antenna. The interrogator consists of two antennas, a radio frequency unit and a processor. When the tag enters a transmission field, it is powered up by the RF (for a passive tag) or its own power source (for an active tag) and starts communicating with the interrogator. When a passive tag is activated, it starts to send all or part of its encoded data, returning to a dormant state when it leaves the field. On the other hand, an active tag is able to receive and store data rather than just send out a fixed signal, and may also have a much greater signal range.

There are four main categories of electronic labelling system. The first of these is *presence sensing*, where the system simply detects that something is there. The second type is *identification*, where as well as sensing that something is there, the system knows exactly what it is. The third type is *transaction*, where the tag contains numerous data fields that are usually updatable, telling the system a variety of information about the tagged object. The final type is the *database tag*, where the tag actually contains instructions for the machines as to what to do to the object.

An example of the application of such a system is in the automotive industry, where car chassis are tracked throughout the system using a passive RF tag containing a vehicle identification number and the various build options. Information regarding specifications

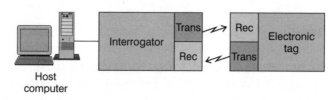

Figure 4.11 *An electronic label system*

such as paint colour is encoded on the tag and provided directly to the automated processes.

4.4.2 Advantages and problems with electronic labels

The main advantage of this form of automatic identification is that the information is capable of evolving and allows the evolution of the object identity during the processing. The reading and writing can be automatic and dynamic, the labels can carry a high density of information and the overall system is reliable, being only slightly sensitive to severe environments.

On the negative side, the technology is expensive, especially in comparison to bar codes, and a very specific system is required to read and write to the labels. There is also the problem of reflection of the radio waves within a manufacturing situation; electromagnetic compatibility (EMC) and contamination can also be a problem.

4.5 Other types of SFDC system

There are a number of other technologies that can be used to form an SFDC system. As with any application, the exact choice of technology is dependent on the specific constraints of the individual environment.

4.5.1 Optical character recognition

In an optical character recognition (OCR) system, the data are read from standard printed text and converted into ASCII characters. This is different from scanning text where the letters are recorded as a bitmap (a digitized representation of a shape) that is not in the same usable format. The only restriction on the data is that they must be printed using a special optical character recognition (OCR) font, of which there are a number of standards, chiefly OCR-A. No special software or printers are required for coding, as OCR fonts are widely available, and this technology has the added advantage that the data can be read by an ordinary operator. The major drawback with OCR is the very high substitution error rate, especially when compared to bar codes, requiring extensive data integrity checks. Manufacturing applications for OCR systems are still very limited, but an example of a system used by Toyota in its supply chain is shown in Figure 4.12, which is based on a CCTV camera as the character reader.

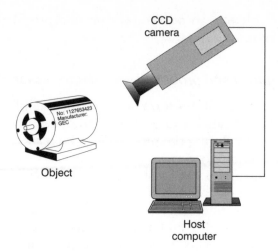

Figure 4.12 *OCR systems used in a supply chain*

4.5.2 Magnetic strips

In this technology, binary encoding is achieved by reversing the polarity of a series of portions of a magnetic band, usually fixed onto a plastic card.

Whilst this system has wide consumer applications, with prime examples being credit and cashpoint cards (Figure 4.13), the use of this technology in manufacturing is still very limited. Drawbacks with this technology include high percentage of bad encoding or misreads, short life of the read/write head, short average life of the card, and the susceptibility to data corruption by exposure to foreign magnetic fields from electric currents or magnetized objects. However, in the correct environment, magnetic strip systems have been successfully implemented.

Figure 4.13 *Magnetic strip cards and card reader*

4.5.3 Direct links to process control devices

Another way in which data can be collected is through direct communication with the programmable logic controller (PLC) or personal computer (PC) used to control an automated process. The controller already logs process parameters such as temperature and pressure and controls machine operations and raw material loading. From this, a whole host of data concerning quality, raw material consumption, machine downtime, etc., is available, and through linking the PLC to a host computer (for example through the RS232 serial port) this can become an integrated part of the shop-floor data collection system. A problem with this method is that one may measure what the controller thinks is happening rather than what is actually happening, unless there are appropriate sensors in place.

4.5.4 Voice recognition systems

Used mainly in the area of quality control, the portable voice recognition data collection system allows the quality control inspector to focus their attention on the part itself without the distraction of needing to type information into a keyboard. Prompts are spoken to the inspector by a voice unit programmed with special application software, and the inspector speaks the answers. Where necessary, the results can be transmitted in real time to a host computer by radio frequency

4.6 The people factor

Whilst the technology is fairly well developed, as with most information systems there are a number of problems with the actual implementation of shop-floor data collection systems in a real situation. One question that must be addressed is ownership of the project; the decision between whether it is the responsibility of manufacturing or data processing must be made. The second problem is that of worker resistance, especially where the system includes the facilities to monitor utilization of the workforce. There is the danger of introducing a 'Big Brother' mentality, resulting in the alienation of workers. Finally, it must be repeated that, in itself, a shop-floor data collection system adds no value to the factory, and if the implementation of such a system is to be justified then the data collected must be acted upon.

4.7 Conclusion

In this chapter, we have made an attempt to define shop-floor data collection as a method of finding out exactly what is happening on the factory floor, with computerized shop floor data collection at its heart today. Various technologies have also been presented. The bar code is certainly the most widespread technology today. However, electronic labels and character and voice recognition methods are used more and more in manufacturing today due to their advantages and technology advances.

4.8 References

Cohen J (1994). *Automatic Identification and Data Collection Systems*, McGraw-Hill, London.

Quinn D (1994). Shop floor data collection – good business uses; *Unpublished proceedings of 1994 Brunel CIM conference*.

Pons J, Chevalier P (1993). *La Logistique Intégrée*, Hermes, Paris.

4.9 Further reading

Monden Y (1998). *Toyota Production System*, Engineering and Management Press, Norcross, Georgia, USA.

Rembold U, Nnaji B.O, Stor A (1993). *Computer Integrated Manufacturing and Engineering*, Addison Wesley, Harlow.

5

Telecommunications

5.1 Introduction

Most of the features of an information system are based on communication technology enabling data transfer from one place to another. It used to be established primarily as direct communication between people, either in direct contact or using mail, telephone or fax. Subsequently telecommunications, defined as the communication of information by electronic means over some distance, has become ever-increasingly important and has also enabled a direct link between computers.

The development of telecommunications is so rapid today that some authors argue that we are currently in the middle of a telecommunications revolution that has two components: rapid changes in the technology of communications and equally important changes in the ownership, control and marketing of telecommunications services (Laudon and Laudon, 1998). It is therefore important to understand capabilities, costs and benefits of alternative communications technologies and how to maximize their benefits for a manufacturing organization.

5.2 Components and functions of a telecommunications system

5.2.1 Telecommunications system components

A telecommunications system is a collection of compatible hardware and software arranged to communicate information from one location to another. Telecommunications systems can transmit text, graphic images, voice, or video information. The major components of a communications system are (Figure 5.1):

- computers to process information;
- terminals or any input/output devices that send or receive data;
- communications channels, the links by which data or voice are transmitted between sending and receiving devices in a network using various media, such as telephone lines, fibre optic cables, coaxial cables and wireless transmission;
- communications processors, such as modems, multiplexers, controllers and front-end processors, which provide support functions for data transmission and reception;
- communications software that controls input and output activities and manages other functions of the communications network (Laudon and Laudon, 1998).

5.2.2 Operation of telecommunications systems

To send and receive information from one place to another, a telecommunications system must perform a number of separate functions that are largely invisible to the people using the system. A telecommunications system transmits information, establishes the interface between the sender and receiver, routes messages along the most efficient paths, performs the elementary processing of the information to ensure that the right message gets to the right

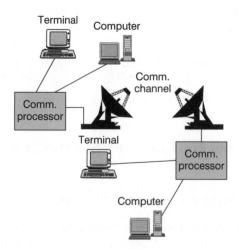

Figure 5.1 *Telecommunications components*

receiver, performs editorial tasks on the data (checking for errors), and converts messages from one speed (speed of a computer) into the speed of a telecommunications line or from one format to another. Finally, the telecommunications systems control the flow of information. Many of these tasks are accomplished by computers.

5.2.3 Telecommunications protocols

A telecommunications network typically contains diverse hardware and software components that need to work together to transmit information (Laudon and Laudon, 1998). Different components in a network can communicate by adhering to a common set of rules that enable them to talk to each other. This set of rules and procedures governing transmission between two points in a network is called a *protocol*. Each device in a network must be able to interpret the other device's protocol.

The principal functions of protocols in a telecommunications network are to identify each device in the telecommunication path, to secure the attention of the other device, to verify correct receipt of the transmitted message, to determine that a message requires retransmission if it is incomplete or has errors, and to perform recovery when errors occur.

5.2.4 Types of signals

Information travels through a telecommunications system in the form of electromagnetic signals (Laudon and Laudon, 1998). Signals are represented in two ways: analogue and digital signals. An analogue signal is represented by a continuous waveform that passes through a communication medium. Analogue signals are generally used to handle voice communication and to reflect variations in pitch. A digital signal is a discrete, rather than a continuous, waveform. It transmits data coded into two discrete states: 0 and 1, which are represented as on-off electrical pulses. Most computers communicate with digital signals, as do many telephone companies and large networks. Since some telephone and network systems still operate using analogue signals, conversion between digital and analogue signals is an important procedure. The device that performs this translation is called a *modem*. A modem translates a digital signal from a computer into an analogue signal to be transmitted over an ordinary telephone line, or other telecommunication media, and vice versa, as shown in Figure 5.2.

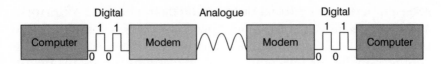

Figure 5.2 *Functions of the modem*

5.2.5 Types of telecommunications channels

Communications channels are the means by which data are transmitted from one device to another (Laudon and Laudon, 1998). A channel can utilize different kinds of telecommunications transmission media: twisted pair, coaxial cable, fibre optics, terrestrial microwave, satellite and wireless transmission.

5.2.5.1 Twisted pair

A twisted pair consists of strands of copper wire twisted in pairs and is the oldest transmission medium. Most of the telephone systems in buildings rely on twisted pair cables. Although it is low in cost, a twisted pair is relatively slow for transmission data, and attempts to increase the speed can cause interference called *crosstalk*. On the other hand, new software and hardware have raised the capacity of existing twisted pair cables up to 10 Mbs (megabits per second), which is often adequate for connecting PCs and other office devices.

5.2.5.2 Coaxial cables

A coaxial cable, similar to that used for cable television, is thickly isolated copper wire, which can transmit a larger volume of data than a twisted pair by reducing crosstalk. It is often used in place of a twisted pair for important links in telecommunications network because it is faster with a speed up to 200 Mbs. However, a coaxial cable is thick, hard to connect and cannot support analogue telephone conversations.

5.2.5.3 Fibre optics

A fibre optic cable consists of one or more strands of clear glass fibre capable of transmitting light signal. Data are transformed into pulses of light, which are sent through the fibre optic cable by a laser device at a rate up to several Gbs (gigabits per second). As such, fibre optic cables are considerably faster that copper wires, lighter and more durable and suitable for systems requiring transfers of a large volume of data. However, fibre optic cables are more difficult to work with, more expensive and harder to install. They are best used as a backbone of a network.

5.2.5.4 Wireless transmission

Wireless transmission sends signals through air or space without any physical tether and acts as an important alternative to twisted pairs, coaxial cables and fibre optic cables. The wireless transmission medium is the electromagnetic spectrum, ranging from radio wave range operating at several kHz up to visible and ultraviolet light operating at 10^{17} Hz. Each frequency range has its own strength and limitations, and these have helped determine the specific function or data communication niche assigned to it.

Microwave systems transmit high-frequency (up to several GHz) radio signals through the atmosphere and are widely used for high-volume, long-distance point-to-point telecommunications. Because microwave signals follow a straight line and do not bend with the curvature of the earth, a long-distance communication requires transmission stations to be positioned about 30 miles apart. This problem can be solved by bouncing microwave signals off satellites, enabling them to serve as relay stations. The satellites move in stationary orbits at about 22 000 miles above the earth and appear to be a very cost-effective method of transmitting large amounts of data.

5.2.6 Characteristics of communication channels

The characteristics of communications channels help to determine the efficiency and capabilities of a telecommunications system (Laudon and Laudon, 1998). These characteristics include the speed of transmission, the direction in which the signals may travel, and the mode of transmission.

5.2.6.1 Transmission speed

The total amount of information that can be transmitted through any telecommunications channel is measured in bps (bits per seconds). Sometimes it is referred to as the *baud rate*. A baud is a binary event representing a signal change from positive to negative or vice versa. The baud rate is not always the same as the bit rate. At higher speed, a single signal can transmit more than one bit at a time, so the bit rate is higher than the baud rate.

The transmission capacity of a telecommunications medium is a function of its frequency. The range of frequencies that can be accommodated on a particular telecommunications channel is called *bandwidth*. The bandwidth is the difference between the highest and the lowest frequencies that can be accommodated on a single channel. Table 5.1 compares the transmission speed and relative cost of the major types of transmission media (Jessup and Valacich, 1999).

5.2.6.2 Transmission modes

There are several conventions for transmitting signals. These conventions are necessary for devices to communicate when a character begins or ends.

Asynchronous transmission (start-stop transmission) transmits one character at a time over a line, each character framed by control bits: a start bit, one or more stop bits and a parity for error checking. Asynchronous transmission is used for low-speed transmission. A serial RS232C communication standard used for serial communication within a PC is a typical example of asynchronous transmission.

Synchronous transmission transmits groups of characters simultaneously, with the beginning and ending of a block of characters determined by timing circuitry of the sending and receiving devices. Synchronous transmission is used for a high volume of data at high speed.

Table 5.1 *Typical speed and costs of telecommunications transmission media*

Medium	Speed	Attenuation	EMI	Cost
Twisted wire	300 bps – 10 Mbs	Moderate to high	High	Low
Microwave	256 Kbs – 100 Mbs	Low	High	Low
Satellite	256 Kbs – 100 Mbs	Low	High	Moderate
Coaxial cable	56 Kbs – 200 Mbs	Moderate	Moderate	High
Fibre optic cable	500 Kbs – 10 Gps	Low	Low	High

5.2.6.3 Transmission direction

Transmission must also consider the direction of data flow over a telecommunications network. In *simplex transmission*, data can travel only in one direction at all times. In *half-duplex transmission*, data can flow two ways but can travel in only one direction at a time. In *full-duplex transmission*, data can travel in both directions simultaneously.

5.3 Types of telecommunications networks

A number of different ways exist to organize telecommunications components to form a network and hence to provide multiple ways of classifying networks. Networks can be classified by their shape or topology. They can also be classified by their geographic scope and type of service they provide.

5.3.1 Network topologies

The three most common network topologies are the *star*, *bus* and *ring* topologies (Laudon and Laudon, 1998).

5.3.1.1 The star network

The star network topology, as shown in Figure 5.3, consists of a central host computer or central processing unit (CPU) connected to a number of smaller computers or terminals and is used primarily in mainframe systems. This topology is useful for applications in which some processing must be centralized and some can be performed locally. One problem with a star topology is its vulnerability. All communication between points must pass through a central computer. If the central computer stops functioning, the network will come to a standstill.

5.3.1.2 The bus network

A bus network (Figure 5.4) links a number of computers by a single circuit made of a twisted pair, coaxial cable or fibre optic cable. All the signals are broadcast in both directions to the entire network, with special software to identify which component to receive the message. There is no central host computer to control the network.

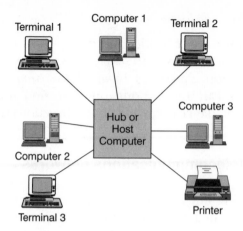

Figure 5.3 *A star network topology*

If one of the computers in the network fails, none of the other components in the network are affected. This topology is commonly used for local area networks (LAN).

5.3.1.3 The ring topology

Like the bus network, the ring network (Figure 5.5) does not rely on a central host computer and will not necessarily break down if one of the component computers malfunctions. Each computer in the network can communicate with other computer, and each processes its own application independently. The connection wire, cable or optical cable forms a closed loop. Data are passed along the ring from one computer to another and always flows in one direction.

The *token ring* network is a variant of the ring network. In the token ring network, all the devices on the network communicate

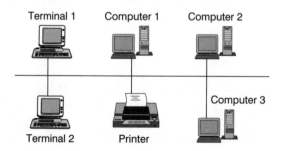

Figure 5.4 *A bus network topology*

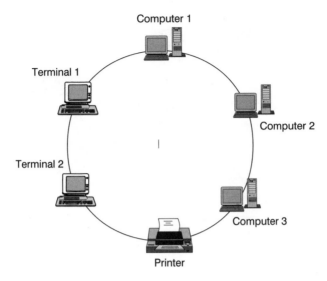

Figure 5.5 *A ring network topology*

using a signal or token. The token is a pre-defined packet of data, which includes data indicating the sender, receiver, and whether the packet is in use. A token moves from device to device in the network, and each device examines it. If the token contains the data for that device, it accepts the data and marks the packet as empty. The token ring configuration is most useful for transmitting a large volume of data between PCs or for transmission between PCs and large computers.

5.3.2 The geographic scope of networks

Networks can be classified by geographic scope into local networks and wide area or global networks (Laudon and Laudon, 1998).

5.3.2.1 Private branch exchange

A private branch exchange (PBX) is a special-purpose computer designed for handling and switching office telephone calls at a company site (Figure 5.6). Today's PBXs can carry both voice and data to create local networks.

While the first PBXs performed limited switching functions, they can now store, transfer, hold and redial telephone calls. They can also be used to switch digital information among computers

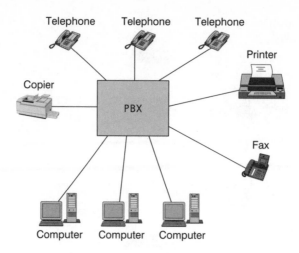

Figure 5.6 *A PBX system*

and office devices. All this activity is possible with digital PBXs connecting smart machines.

The advantage of digital PBXs over local networking options is that they utilize existing telephone lines and do not require special wiring. A simple plug-and-play principle can be used in most cases. It is only the mainframe computer that requires coaxial or fibre optic networking to wide area network. The networks are generally simple to maintain and the company does not need specialized expertise to manage them. The primary disadvantage of PBXs is that they are limited to telephone lines and they cannot easily handle a very large volume of data.

5.3.2.2 Local area networks

A local area network (LAN) encompasses a limited distance, usually one building or several buildings in close proximity. Most LANs are limited to a single company site and are widely used to link PCs. LANs require their own communication channels.

LANs generally have a higher transmission rate than PBXs, up to several Mbs, and they are required for applications with a high volume of data. LANs are capable of transferring video signals and graphics.

LANs are totally controlled, maintained and operated by end-users. This provides a high level of flexibility to the company, but it also means that the user must know a great deal about telecommunications applications and networking. The most common application of a LAN is networking of PCs within a building or office to share information and expensive peripheral devices (printers, storage tapes, etc.)

Figure 5.7 illustrates a LAN. The server acts as a librarian, storing various programs and data files for network users. It also determines who gets access to what and in what sequence. Servers are usually powerful PCs or workstations with large hard disk capacity. Specialized computers are now available to act as LAN servers, containing network operating systems that manage the whole network.The network gateway connects the LAN to public networks, such as telephone networks or other corporate networks, for exchange of information with external world. A gateway is generally a communications processor that can connect dissimilar networks by translating from one set of protocols to another.

LAN technology consists of cabling (twisted pair, coaxial or fibre optic cables), network interface cards, and software to control LAN activities. Sometimes, wireless communication is used to connect remote sites. The LAN interface cards specify the data transmission rate, the size of message units and network topology. The most popular Ethernet LAN uses a bus topology.

Although LAN networks generally provide a great deal of flexibility in connecting different devices within a company, their major disadvantage is in the fact that they are more expensive than PBXs and require new wiring each time the LAN is moved. LAN also requires specially trained staff to manage and run them.

5.3.2.3 Wide area networks

A wide area network (WAN) spans broad geographic distances, ranging from a few miles to across entire continents. WANs may consist of a combination of switched and dedicated lines, microwave and satellite communications. Switched lines are telephone lines that a person can access from a terminal to transmit data to another computer, the call being routed or switched through paths to the designated destination. Dedicated, or *non-switched* lines, are continuously available for transmission. The lines can be leased or purchased from common carriers or private communications media vendors. Dedicated lines are often conditioned to transmit data at higher speeds than switched lines and are more appropriate for

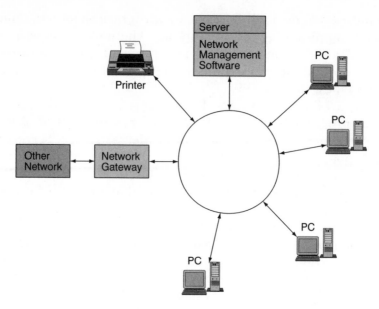

Figure 5.7 *A local area network*

high-volume transmissions. Switched lines, on the other hand, are less expensive and more appropriate for low-volume applications that require only occasional transmission. Individual business firms may maintain their own WANs.

5.3.2.4 Value-added networks

Value-added networks (VANs) are an alternative to firms designing and managing their own networks. VANs are private, multi-path, data only, third party-managed networks that can provide economies in the cost of service and in network management because they are used by multiple organizations. The VAN is set up by a firm that is in charge of managing the network. The firm sells subscriptions to other firms wishing to use the network. The fee is usually determined according to the amount of data transmitted over the network. The network may utilize twisted pair lines, satellite links and other communications channels leased by the value-added carriers.

The term *value added* refers to the extra value added to communications by the telecommunications and computing services these networks provide to clients. Customers do not have to invest in network equipment and software. Subscribers may achieve savings in line charges and transmission costs because the costs of using the network are shared among many users.

5.4 Telecommunications for competitive advantage

Telecommunications has helped to eliminate barriers of geography and time, enabling organizations to accelerate the pace of production, to speed decision making, to forge new products, to move into new markets and to create new relationships with customers and suppliers. Firms that fail to consider telecommunications in their strategic plans will fall behind (Laudon and Laudon, 1998).

5.4.1 Telecommunications applications

Some of the leading telecommunications applications for communication, co-ordination and speeding the flow of transactions, messages and information through a business firm are electronic mail, voice mail, facsimile machines, digital information services, teleconferencing, videoconferencing, electronic data interchange and groupware (Jessup and Valacich, 1999).

5.4.1.1 Electronic mail

Electronic mail, or e-mail, is the computer-to-computer exchange of messages. A person can use a PC attached to a modem or a networked terminal to send notes and even lengthy documents just by typing the name or computer address of the recipient. Many organizations operate their own internal e-mail systems, but communications companies such as MCI and American Telephone and Telegraph (AT&T) offer this service, as do commercial on-line information services such as America Online, CompuServe, Prodigy and public networks on the Internet. E-mail eliminates telephone tag and costly long-distance telephone charges.

E-mail systems present security problems because without adequate protection, electronic eavesdroppers can read the e-mail as it moves through the network.

5.4.1.2 Voice mail

Voice mail digitizes the spoken message of the sender, transmits it over a network, and stores the message on the disk for later retrieval. When the recipient is ready to listen, the message is reconverted to audio form. Various store-and-forward capabilities notify recipients that messages are waiting. Recipients have the option of saving messages for later use, deleting them, or routing them to other partners.

5.4.1.3 Facsimile machines

Facsimile machines can transmit documents containing both text and graphics over an ordinary telephone line. A sending fax machine scans and digitizes the document image. The digitized document is then transmitted over the network and reproduced in a hard copy or displayed on the screen by a receiving fax machine. The process results in a duplicate, or facsimile, of the original.

5.4.1.4 Electronic data interchange

Electronic data interchange (EDI) is an inter-organizational communication system forming a direct link from the computer systems of one organization to the computer systems of another, allowing information and instructions to pass between the two companies without the need for human intervention. EDI is yet another service that heavily depends on the communications networks. Still, it is fundamentally different from other communications media, such as faxes, e-mail and paper mail, in that the information sent is immediately usable by the receiving system. The definition adopted by the National Computing Centre (NCC) is that EDI is defined as *the transfer of structured data, by agreed message standards, from one computer system to another, by electronic means.*

A simpler definition is that EDI is the practical method of engaging in paper-less trading, allowing commerce to take place without the need for large amounts of administrative paperwork.

5.4.1.5 Digital information services

Powerful and far-reaching digital electronic services now enable networked PCs and workstation users to obtain information from outside the firm instantaneously without leaving their desks. Stock prices, historical references to periodicals, industrial supplies catalogues, legal research, new articles, reference works, weather forecasts and travel information are just some of the electronic databases that can be accessed on-line. Many of these services have capabilities for e-mail, electronic bulletin boards, and for on-line discussion groups, shopping and travel reservations. Table 5.2 describes the leading commercial digital information services. Capabilities of the Internet can also be considered as a form of digital information service.

5.5 Management issues and decisions

The starting point for rational planning of telecommunications is to forget about the features of systems and instead try to understand requirements of the organization. Cutting costs and installing advanced systems for their own sake is rarely a sufficient reason to justify large telecommunications projects.

5.5.1 The telecommunications plan

Although it is very difficult to generalize the way of planning telecommunications systems in any company, there are verified steps to implementing a strategic telecommunications plan (Laudon and Laudon, 1998).

First, start with an audit of the communications functions in the company. What are voice, data, video, equipment, staffing, and management capabilities? Then identify priorities for improvement.

Second, the long-range business plans of the firm must be known. The plan should include an analysis of precisely how telecommunications will contribute to the specific five-year goals of the firm.

Table 5.2 *Commercial digital information services*

Provider	Type of Service
America Online	General interest/business information
CompuServe	General interest/business information
Prodigy	General interest/business information
Microsoft Network	General interest/business information
Dow Jones News Retrieval	Business/financial information
Quotron	Financial information
Dialog	Business/scientific/technical information
Lexis	Legal research
Nexis	News/business information

Third, identify critical areas where telecommunications currently does or can have the potential to make a large difference in performance. In insurance, for instance, these may be systems that give field representatives quick access to policy and rate information.

5.5.2 Implementing the plan

Once an organization has developed a business telecommunications plan, it must determine the initial scope of the telecommunications project. Managers should take eight factors into account when choosing a telecommunications network:

- the first and most important factor is distance;
- the range of services the network must support: e-mail, EDI, voice mail, videoconferencing, imaging, etc.;
- security. The most secure form of telecommunications is through dedicated leased lines;
- whether a multiple access is required throughout the organization or whether it can be limited to one or two nodes within the organization;
- utilization, which is the most difficult factor to judge. There are two factors related to utilization: the frequency and the volume of communications;
- the cost of each telecommunications option including the cost for development, operations, maintenance, expansion and overhead. Which are fixed and which are variable costs? Identify any hidden cost;
- the difficulties in installing the telecommunications system;
- consider how much connectivity would be required to make all the components in a network communicate with each other (different standards for hardware and software).

5.6 Conclusion

In this chapter, telecommunications has been defined as data transfer over a distance using either analogue or digital electrical signals. It can be established using twisted pair wires, coaxial cables, fibre optical cables or wireless transmission. Each of these media has its own technical advantages and disadvantages. Additionally, telecommunication components can be connected using three network topologies: star, bus and ring topologies. In a geographic sense, telecommunication networks can be classified as PBXs, LANs, WANs and VANs. Various applications of telecommunication systems have also been discussed.

5.7 References

Jessup L.M, Valacich J.S (1999). Information Systems Foundations, Education and Training.

Laudon K.C, Laudon J.P (1998). *Management Information Systems – New Approaches to Organisation & Technology*, Prentice Hall, Upper Saddle River, New Jersey, USA.

5.8 Further reading

Alter S (1999). *Information Systems: A Management Perspective*, Addison Wesley, Harlow.

Curtis G (1998). *Business Information Systems – Analysis, Design and Practice*, Addison Wesley, Harlow.

Forouzan B.A (2000). *Data Communications and Networking*, McGraw-Hill International Edition, London.

Peppard J (1993). *IT Strategy for Business*, Pitman Publishing, London.

Sokol P.K (1989). *EDI: The Competitive Edge*, Intertext Publications, New York.

6

Electronic Commerce

6.1 Introduction

The focus of this section is on how the Internet works and how companies use it to streamline operations, sell products, provide customer support, or connect to suppliers. Broadly speaking, *electronic commerce* refers to use of the Internet to support day-to-day business activities, which demonstrates the fact that the influence of the Internet is not limited to technical people only. Marketing professionals focus on how to use the Internet for selling products and providing customer service. Finance professionals use the Internet to get real-time market updates. Managers must deal with employees working from remote locations using the Internet as the primary means to communicate with colleagues.

6.2 The Internet

The name *Internet* is derived from the concept of inter-networking, that is, connecting host computers and their networks to form a larger, global network (Jessup and Valacich, 1999). That is essentially what the Internet is: a large worldwide network of networks that use a common protocol to communicate with each other. The interconnected networks include UNIX, VAX, IBM, Novell, Apple and many other network and computer types. The networks that constitute the Internet are each developed and maintained by different organizations, ranging from government agencies and educational institutions to private businesses and large commercial services such as American Online (AOL) and CompuServe. No single person or organization owns or maintains the Internet.

The Internet enables companies, groups, institutions and individuals to share a wide range of data, including text, video, audio, graphics, databases and other media types. The scientific and academic communities have used it for many years for information-sharing and research. Recently, the largest growth segment of the Internet has been the business sector.

6.2.1 Operation of the Internet

6.2.1.1 Packet-switching technology

The Internet follows a hierarchical structure. High-speed central networks, called *backbones*, are highways enabling traffic from mid-level networks. It relys on packet-switching technology to deliver data and information across networks. Data are transmitted in small chunks concurrently (Figure 6.1). To minimize delays, network technologies limit the amount of data that a computer can transfer on each turn. Each packet being sent across the network must be labelled with a header. The header contains the network address of the source (*sending* computer) and the network address of the destination (*receiving* computer). Each computer attached to a network has a unique network address. As packets are sent, network hardware detects whether a particular packet is destined for a local machine.

6.2.1.2 Connecting independent networks

The Internet uses special-purpose computers called *routers* to interconnect independent networks, as shown in an example illustrated in Figure 6.2 (Jessup and Valacich, 1999). However, routers do not use conventional software nor are they used to run applications.

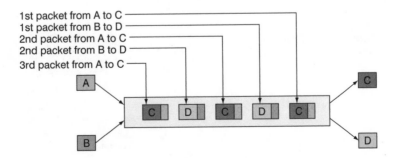

Figure 6.1 *Packet switching*

Their only job is to interconnect networks and forward data packages from one network to another. Essentially, routers are the means by which thousands of LANs and WANs are interconnected. Typically, LANs are connected to backbone WANs. A backbone network manages the bulk of traffic and uses higher-speed protocols that the individual LAN segments.

To gain access to the Internet, an organization generally connects a router between one of its own networks and the closest Internet site. An individual, however, typically uses a modem and standard telephone dial-up service to connect to a computer or network that is connected to the Internet. Many commercially available servers offer this connection free of charge.

6.2.1.3 The Internet communication protocol

The common language, or protocol, that interconnects different network and computer types to the Internet is called the *transmission control protocol/internet protocol* (TCP/IP) (Laudon and Laudon, 1998). The first part, TCP, breaks information into small chunks (data packets) that are transferred from one computer to another. Typically, a data packet contains several hundred characters. The IP part of the protocol attaches the destination address to the packet. Packets travel independently to their destination, sometimes following different paths and arriving out of order. The receiving computer reassembles the packets based on their identification and

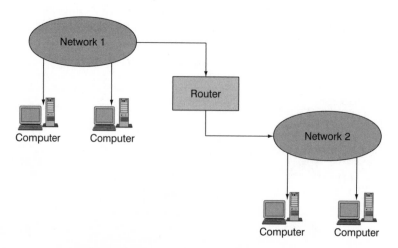

Figure 6.2 *Routers connect networks*

sequencing information. Together, TCP and IP provide a reliable and efficient way to send data across the Internet.

Every computer that uses the Internet must run IP software. The IP defines how a data packet must be formed and how a router must forward each packet. A data packet that conforms to the IP specifications is called an *IF datagram*. Every computer, including routers, is assigned a unique IP address. Every user also must have a different name. The combination of the user names and IP addresses compose the Internet address. The IP address and the user name are separated by the '@' symbol.

6.2.1.4 Internet services

Many tools and applications exist for using the Internet. These applications follow the client/server model. Client programs run on the local computer, such as a PC, and facilitate information access by doing the behind-the-scenes work of opening connections to distant computers, sending requests and receiving and displaying results. The server software runs on computers that provide information. A server is usually a very powerful computer capable of simultaneously handling information requests from many clients. A collection of the Internet tools follows this approach and enables users to exchange messages, share information, or connect to remote computer. These tools are summarized in Table 6.1.

6.2.1.5 The World Wide Web

The World Wide Web (WWW), or simply the Web, is the most powerful use of the Internet. More than likely, you have browsed the Web using Netscape Navigator, Microsoft Internet Explorer, or some other popular browser, as shown in Figure 6.3.

Although Gopher can tie together text documents, Telnet sessions, sounds, graphics, file transfers and more in a menu-driven format, the Web takes Gopher one step further by using hypertext. A hypertext document not only contains the information, but also references or links to other documents that contain related information. The standard method of formatting Web pages is to use hypertext mark-up language, or HTML. HTML is much like a programming language that operates through a series of codes placed within a text document and result, with the help of a Web browser, in a formatted Web page. Additionally, each Web page again has a unique address called a *uniform resource locator* (URL).

Table 6.1 *Internet tools and their description*

Internet Tool	Description
E-mail	Enables users to send messages to each other
Telnet	Enables users to connect, or log in, to any computer on the Internet
File transfer protocol (FTP)	Enables users to connect to a remote computer solely for the purpose of transferring files: either uploading (sending to the remote machine) or downloading (getting back from the remote machine) files and data
Listserv, short for 'mailing list server'	Enables groups of people with a common interest to send messages to each other. Interested people subscribe to a discussion group, which is essentially a mailing list. When a subscriber sends a message to the list, the message is sent to all other subscribers
Usenet	Enables groups of people with a common interest to send messages or other binary information to each other. Unlike listserv, Usenet has no master list of subscribers. Rather, anyone with access to Usenet may use a new reader program to post and read articles from the group
Archie	Enables users to search FTP sites for their contents. For example, you might be looking for a particular file, perhaps software application or game. You would use Archie to search FTP sites. Using the results of the Archie search, you can determine which FTP site has the desired files, and then use FTP to download them
WAIS (Wide Area Information Server)	Enables users to locate information by indexing electronic data using standard keywords
Gopher	A text-based, menu-driven interface that enables users to access a large number of varied Internet resources as if they were in folders and menus on their own computers. Menu choices on a Gopher server include text files, graphic images, sounds, software, or even another menu

6.2.2 Technologies enabling Internet communication

As described earlier in this section, many technologies are used to create the Internet, in particular, advances in computer networks and telecommunications (Jessup and Valacich, 1999). This section briefly describes three important technologies: integrated services digital network (ISDN), T1 service and asynchronous transfer mode (ATM).

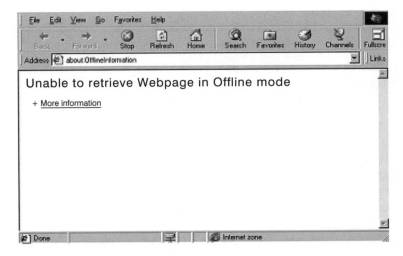

Figure 6.3 *Microsoft Internet Explorer*

6.2.2.1 Integrated service digital network

ISDN is a worldwide standard for digital communications, and is intended to replace all current analogue systems such as telephone connections. Its great strength is that it can use existing twisted pair telephone wires for high-speed data services. As ISDN is purely a digital network, the need for a modem will vanish.

For reasons of consistency, Table 6.2 summarizes all the important standards to achieve connectivity.

6.2.2.2 T1 service

A T1 line is a dedicated digital transmission line that can carry 1.544 Mbs of information over a long distance, up to hundreds of thousands of miles. They were initially developed for enabling a proper EDI transmission, but today they are mainly used to connect major telephone exchanges. T3 lines are used for faster links, as shown in Table 6.3.

6.2.2.3 Asynchronous transfer mode

An ATM is a method of transferring voice, video and data over high-speed LANs. The transfer speed is up to 2.2 Gbs. An ATM uses a form of packet transmission in which data are fixed in length, in a 53-byte cell.

Table 6.2 *Standards for achieving connectivity*

Area	Standard or Reference Model	Description
Networking	TCP/IP, Operating systems interconnect (OSI), Systems network architecture (SNA)	Computer-to-computer communications
Digital public switched network transmission	ISDN	Transmission of voice, video and data over public telephone lines
Optical fibre transmission	FDDI	100 Mbs data transmission over dual optical fibre ring
E-mail	X.400	Permits e-mail system operating on different hardware to communicate
Packet-switching	X.25	Permits different international and national networks to communicate
EDI	X.12 Edifact (Europe)	Standardized data-only transaction format
Graphical user interface	X Windows	High-level graphics description for standardized Windows management
Operating system	UNIX	Software portable to different hardware platforms

Table 6.3 *Capacity of communication lines*

Type of Line	Data Rate (Mbs)	Equivalent Number of Voice Lines
T1	1.544	24
T3	44.736	672

6.3 Electronic commerce

Electronic commerce is the on-line exchange of goods, services and money within companies, between companies and between companies and their customers. The allure of electronic commerce is that it has no geographical or time limit.

Electronic commerce now takes place in a number of different ways and on a number of different technology platforms. It can be argued that the beginning of electronic commerce was with the introduction of EDI. However, the current trend in business today is to use the public Internet. It is estimated that, in the UK, companies buy about £300 billion worth of goods and services via traditional EDI networks each year. Estimates of the volume of Internet-based transactions for 1995 ranged from several hundred million pounds to £3 billion, with an increase to £100 billion in 1997.

6.3.1 A model of electronic commerce

Company web sites range from the passive to the active (Jessup and Valacich, 1999). At one extreme are the relatively simple, passive web sites that provide only product information and the company address and phone number, much like a traditional brochure. At the other extreme are relatively sophisticated, active web sites that enable customers to see products, services and related real-time information, and actually conduct purchase transactions on-line. A typical four-stage development of a web site is shown in Table 6.4 (Laudon and Laudon, 1998).

Figure 6.4 shows a model of electronic commerce with five phases: information gathering, ordering, payment, fulfilment and service and support. Companies now advertise their offerings to prospective customers through the Internet, e-mail and the web. Customers can order and pay for products and services on-line. If the product or service can be digitized, it can be delivered on-line, as with information, videos and software products. Finally, some innovative companies are finding ways to provide on-line service and support after the sale. This can be done with

Table 6.4 *A typical evaluation path of web site development*

Stage	Activities
Corporate image and product information	Register a domain name and create pages providing contact information about the company and its products
Information collection and market research	Create forms with which customers can register their identity on-line and assign account numbers; create marketing research form
Customer support and service	Link web site to tracking database: customers can enter a package number and view up-to-date information on the stage of their orders and deliveries
Transactions	Enable customers to request delivery and arrange payment options

support documentation, videoconferencing with helpful product support staff or, as with computer hardware, the vendor can diagnose faults through the network.

The trend has been recently toward Internet-based electronic commerce. Figure 6.5 shows the possible modes of electronic commerce using the Internet. The term used to describe transactions between individuals and firms is *Internet-based electronic commerce*. *Intranet*, on the other hand, refers to the use of the Internet within the same business. *Extranet* refers to the use of the Internet between companies.

Potential electronic commerce income does not come from on-line sales transactions only. A great deal of money has been made by companies advertising on their web sites. Many companies are now beginning to collect subscription fees from web surfers for valuable on-line newspapers, magazines and other resources. Additionally, high interest in having web sites has created a demand for skilled people to design and manage web sites – *webmasters*.

6.3.2 Exploiting Internet-based electronic commerce

There are two old rules of commerce (Jessup and Valacich, 1999):

Rule #1:   Offer something of value.
Rule #2:   Offer products and services at a fair price.

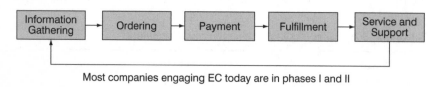

Most companies engaging EC today are in phases I and II

Figure 6.4 *A model of electronic commerce*

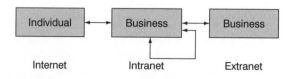

Figure 6.5 *The possible modes of electronic commerce*

These old rules apply to EC too. However, some new rules related to EC have been created. These are:

Electronic commerce Rule #1:   The web site must be aesthetically pleasing.

Electronic commerce Rule #2:   The web site must be easy to use and fast.

Electronic commerce Rule #3:   The web site must motivate people to visit, to stay and to return.

Electronic commerce Rule #4:   Advertise your presence on the web.

Electronic commerce Rule #5:   Learn from your web site.

These rules mean that the companies that were traditionally successful in the old markets will not necessarily dominate the new electronic markets. Although the above rules are rather self-explanatory, it is suffice to stress some of the associated issues.

The web sites must be fast to download. Early research suggests that the average length of time a web surfer will wait for a web page to download is only a couple of seconds. The web site has to motivate the user. The most popular web sites, for instance, are Microsoft web pages, primarily because of the free software the user can download. The possibility of interacting with other users can also be attractive.

Web sites employ more of a *pull* than *push* marketing strategy. A company must pull visitors to its site and away from thousands of similar sites. Smart companies are good at advertising their web sites, usually in two ways:

- include the web site address on all company materials, from business cards and letterheads to advertising copy;
- advertise the web site on other popular web sites, such as the ESPN SportZone or one of the more popular search engines.

Learning from the web site is an important task. There are commercially available software packages to analyse the usage of a web site, including the number of visitors, and how people navigate through the site. This can indicate the amount of interest for the products and, in particular, which product is higher on the ladder.

6.3.3 Intranets as a first step into electronic commerce

Many firms have decided to utilize the Internet to support internal operations. They literally block out unwanted users on the Internet from entering their web pages and servers by using password protection, firewalls and limiting the access through the router. Firewalls are hardware and software placed between an organization's internal network and an external network to prevent outsiders from invading private networks.

Some common uses of intranets include providing access to on-line internal phone books, procedure manuals and training materials, enabling employees to check inventories and order supplies and to create personal benefit plans. These could be easily extended to a centralized database of all administrative documents with general (or limited) but instant access.

Intranets are popular for several reasons. First, an intranet gives an organization a relatively inexpensive, easy, quick international telecommunications infrastructure. Second, intranets can help companies improve product and service quality while decreasing cost and cycle time. Third, the use of an intranet helps organizations learn more about the Internet before interacting with real customers. Intranets are often a wise intermediate step in preparing for full-blown electronic commerce.

6.3.4 From EDI to Internet-based extranets

In addition to intranets, many companies are also using the Internet to support their dealings with other companies. The business-to-business commerce takes many forms:

- manufacturer's order of materials from a supplier;
- travel arrangements with travel agencies;
- collaboration on advertising copy with advertising companies;
- person connected to another company computer or one company computer working with another company computer with no people involved;
- a complete vertical link of buyer/supplier/bank.

As mentioned earlier, the trend currently is toward Internet-based extranets rather than the more traditional EDI using proprietary networks. This shift makes sense:

- the Internet provides a global telecommunications infrastructure for which it is relatively easy and inexpensive to develop applications;
- everyone uses documents based on HTML format, and packet switching via the TCP/IP network protocol;

- users have lots of good choices for browsers, HTML editors, web server software, and other related tools;
- connection to the Internet is rather inexpensive through local and national Internet service providers;
- security mechanisms on the Internet are improving every day.

6.4 References

Jessup L.M, Valacich J.S (1999). Information Systems Foundations, Education and Training.
Laudon K.C, Laudon J.P (1998). *Management Information Systems – New Approaches to Organisation & Technology*, Prentice Hall, Upper Saddle River, New Jersey, USA.

6.5 Further reading

Peppard, J (1993). *IT Strategy for Business*, Pitman Publishing, London.
Sokol, P.K (1989). *EDI: The Competitive Edge*, Intertext Publications, New York.

Strategic Implications of MIDS

7.1 Introduction

Despite the fact that computer systems have been used in business for a number of decades, it is only during the last few years that the focus has started to shift from the actual technology to the integration of information systems into businesses in order to enhance business strategy.

The first subject dealt with in this chapter looks at the evolution of an information system within an organization, examining the attitudes towards management and applications of computer systems since their introduction. Following this, a number of strategic models for examining the potential applications of information systems within an organization will be detailed, before finally moving on to examine some of the practical issues relating to evolving a successful information systems strategy. One important definition for this module is drawing the difference between information technology and information systems. Although this has been detailed in Chapter 1, suffice it to say that IT refers to the actual technologies used, whilst IS refers to the integration of the technology into the organization as a system.

7.2 Concept of strategic information systems

Strategic information systems change the goals, business processes, products, services, or environmental relationships of organizations to help them gain an edge over competitors. Systems that have these effects may even change the business of organizations.

Strategic information systems often change the organization as well as its products, services, and internal processes, driving the organization into new behaviour patterns. The organization may need to change its internal operations to take advantage of the new information technology. Such changes often require new managers, a new workforce, and a much closer relationship with customers and suppliers.

7.3 The evolution of IS/IT

Behind the growing strategic uses of information systems is the changing conception of the role of information in organizations. Organizations now consider information to be a resource, much like capital or labour. This was not always the case.

In the past, information was often considered a necessary evil associated with the bureaucracy of designing, manufacturing and distributing a product or service (see Table 7.1). Information systems of the 1950s focused on reducing the cost of routine paper processing, especially in accounting. By the 1960s, organizations started viewing information differently, recognizing that information could be used for general management support. Any information system of the 1960s and 1970s was frequently called a management information system (MIS) and was thought of as an information factory churning out reports on weekly production, monthly financial information, inventory, accounts, etc. Additionally, the computers were introduced in the manufacturing process to improve operational efficiency, thus this period is commonly known as the *automation period*. In the 1970s and early 1980s, information, as well as the systems that collected, stored and processed it, were seen as providing fine-tuned, special-purpose, customized management control over the organization. The information systems that emerged during this period were decision-support systems and executive support systems. Their purpose was to improve and speed up the decision-making process of specific managers and executives in a broad range of problems, primarily at a tactical level. As a consequence of realizing the importance of information, this period is commonly called the *information period*. By the mid-1980s, the conception of information changed again. Information began to be viewed as a strategic asset or resource: it began to be seen as a source of strategic advantage, or a weapon to defeat and frustrate competition. These changing conceptions of information reflected advances in strategic planning and theory, hence this period is known as the *period of transformation*.

Moving toward the year 2000 and beyond, the arrival of high-power computing and universal networking via the Internet once again changed the business and manufacturing concept of information. Now the information is viewed as the very foundation of business process, products and services. Firms are seen primarily as composed of knowledge assets that, if properly shared with vendors, customers, and employees, can become the foundation for sustained success and relative competitive advantage.

7.4 Use of IS for competitive advantage

Companies use information technology at three different levels of strategy: the business, the firm and the industry level (see Table 7.2). Generally, there is no single all-encompassing strategic system,

Table 7.1 *Changing concept of information system*

Time Period	Conception of Information	Information System	Purpose
1950–1960	**Necessary evil,** bureaucratic requirement, a paper dragon	**Electronic accounting machine**	Speed accounting and paper processing
1960–1970	General purpose support	Management information system (MIS), information factory	Speed general reporting requirements
1970–1980	Customized management control	Decision support system, executive support	Improve and customize decision-making
1985–2000	Strategic resource, competitive advantage,business foundation	Strategic system	Promote survival and prosperity of the organization

but instead a number of systems operating at different levels. For each level of business strategy there are strategic uses of systems, and for each level of business strategy there is an appropriate model used for analysis.

7.4.1 The value chain

7.4.1.1 The basic model

The competitive advantage cannot be understood by looking at a firm as a whole. It does, however, stem from the many discrete activities a firm performs in designing, producing, marketing, delivering and supporting its products (Peppard, 1993). The combinations of discrete activities, performed by the company to add value to a product, are known as *value activities*. Their combination in sequential order forms the value chain for the company, which is proposed as the basic tool for analysing the sources of competitive advantage. In order to be profitable, the cost of performing the value activities must not exceed the value created, and in order to provide a competitive advantage the firm must:

- perform these activities at a lower cost than competitors; or
- perform them in such a way that permits differentiation at a premium price.

The generic value chain is shown in Figure 7.1.

Table 7.2 *Strategy levels of IT*

	Strategies	**Models**	**IT Techniques**
Industry	Co-operation vs competition licensing standards	Competitive forces models Network economics	Electronic transaction Communication networks Inter-organizational systems Information partnership
Firm	Synergy Core competencies	Core competency	Knowledge systems Organization-wide systems
Business	Low cost Differentiation Scope	Value chain analysis	Data mining IT-based products/services Inter-organizational systems Supply chain management Efficiency customer response

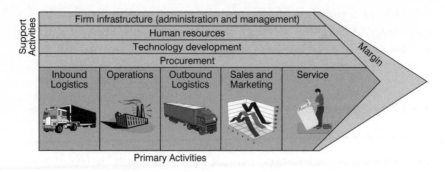

Figure 7.1 *Activities of the value chain*

As seen, the activities in the value chain can be classified into *primary activities* and *support activities*.

Primary activities are most directly related to the production and distribution and they add value to the product or service directly, involving the physical creation of a product and its sale and distribution to the buyer, as well as after-sales service.

Support activities are those that incur costs whilst not directly adding value to the product. They are, however, essential to support the primary activities and each other by providing purchased inputs, technology, human resources and various firm-wide functions.

The definitions of the activities shown in Figure 7.1 are as follows:

Inbound logistics:	inbound activities to receive, store and distribute inputs to the product, such as material handling, inventory control, warehousing and contacts with suppliers.
Operations:	production activities to create the product such as machining, packaging, printing and testing.
Outbound logistics:	outbound activities to store and distribute the product to customers, including warehousing, order processing and vehicle scheduling.
Marketing and sales:	activities associated with providing a means by which buyers can purchase the product or be induced to do so (advertising, selling, pricing, merchandising, promotion).
Service:	after-sales activities for providing service or maintaining product value, including installation, repair, parts and training.

Procurement:	purchasing inputs.
Technology development:	not just machines and processes but also expertise, procedures, and systems.
Human resources:	activities involved in the recruiting, training, development, and remuneration of staff.
Firm infrastructure:	general management, finance, planning, quality assurance. The infrastructure supports the whole value chain.

An example of the value chain for a manufacturing company is shown in Figure 7.2.

Following on from the overall view, it is now possible to look at how the value chain concept can be used in the formulation of IT/IS strategy.

7.4.1.2 Applying the value chain to IT/IS strategy

Information plays a very important part in the value chain. For example, sales orders, product specifications and quality control systems are all types of information that are utilized in the production system. By examining the different components and the information flows involved, it is possible to pose questions as to how IT/IS can transform the value chain. For example, can IT/IS:

- contribute to performing an activity more quickly or more efficiently or perhaps at a lower cost than before;
- improve information flow through the primary activities;
- be used to affect how support activities assist primary activities (e.g., finance, budgetary control)?

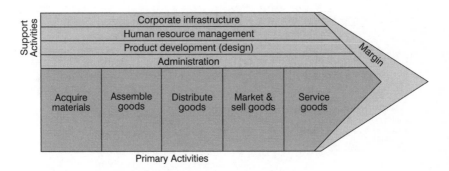

Figure 7.2 *The manufacturing value chain*

The main benefit of value chain analysis is that it identifies the main information needs and flows that reflect what the business actually does (or would like to do), as opposed to how it is organized to do it (the organization structure). In addition, it moves the focus from IT/IS as a cost to a value-adding process.

Various examples of strategic information systems for the primary and support activities of a manufacturing firm that would add a margin value to a firm's products or services are shown in Figure 7.3.

The role of information technology at the business level is to help the firm reduce costs, differentiate products, and serve new markets. Table 7.3 shows some leading examples of how firms use IT to lower costs, differentiate, and change the scope of competition.

7.4.2 Five competitive forces model

One of the basic models in the strategic toolkit is the 'five forces' model of industry. Based on the principle that each industry is characterized by five competitive forces, by examining these forces a number of strategic options for developing or enhancing a company's competitive advantage within that industry can be generated. The basic model is shown in Figure 7.4. It is important for this form of analysis that the firm is analysed as strategic business units, which are subdivisions of the firm's entire business, each competing in their own, well-defined markets. The simplest illustration of this is to consider a multinational company with a range of interests in different industries. It would not be appropriate to examine the whole corporation's competitive position,

Figure 7.3 *Various examples of strategic information systems*

Table 7.3 *New products and services based on information technology*

New Products or Service	Underlying Technology
On-line banking	Private communication networks, Internet
Cash management accounts	Corporate-wide customer account systems
Derivative investments (options, futures, complex variations)	Management and trader workstations, mainframe transaction systems
Global and national airline, hotel, and reservation system	Worldwide telecommunication-based auto reservation systems
Overnight package delivery	Nationwide package training systems
Mail-order retailing	Corporate customer databases
Voice mail systems (call services)	Company and network-wide digital communication systems
Automatic teller machines	Customer account systems
Micro-customized clothing	CAD and CAM systems

where they may be operating in markets as diverse as consumer durables and industrial products.

The five forces shown in Figure 7.4 are explained as follows.

The first is the bargaining power of the buyers. In industries with only a small number of customers, they can exert a strong influence over the suppliers, causing them to undercut each other in order to get business. They can also use their power in other ways, for example, by demanding credit or improved quality of product. However, companies can influence this force, for example, by building in switching costs, i.e., costs associated with switching suppliers or using substitute products, hence tying the customer to them.

The next force is the bargaining power of suppliers, which is similar to that of buyers but seen from the opposite perspective. If the suppliers are in a strong position, then they can increase a firm's costs by extracting a higher price for goods and services. This time, the company wishes to minimize switching costs and have a wide choice of suppliers. Another useful tool is the threat of backward integration, where a company secures inputs through their own resources rather than relying on suppliers.

The threat of new entrants also influences the competitiveness of an industry. New competitors bring additional capacity to the market, often with substantial resources to compete with. Unless the market is expanding faster than new entrants are arriving, there will be eating away of profits so profitability will decline. The main

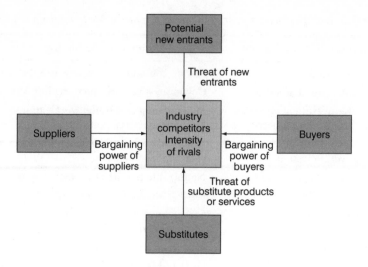

Figure 7.4 *The five competitive forces*

components of an industry affecting this force are called *entry barriers*, the most common of these being economies of scale, brand preference and customer loyalty, high capital requirements, access to distribution channels and legislation and customer policy.

The influence of substitute products is also an important industry-shaping force. The propensity for buyers to switch to an alternative product is affected by both the relative price and performance of the substitute product, and the switching costs associated with making the transition. High switching costs require a substitute with a high price/performance benefit to be a threat. Substitution also allows a company to expand their own market by offering its products as substitutes to another buyer.

The final competitive force, and arguably the most powerful, is that of rivalry between competitors within the industry. Each firm attempts to employ competitive strategy to improve their position and competitive edge, by offering buyers something that their competitors cannot duplicate easily or cheaply.

7.4.2.1 The effect of IT/IS on the five forces

Utilizing the above analysis of an industry, it is possible to examine how the implementation of IT/IS can shape the forces characterizing an industry. The strategic questions are shown in Figure 7.5 (McLaughlin, 1994).

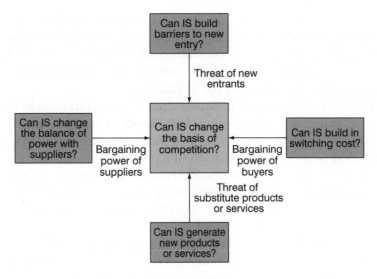

Figure 7.5 *Strategic IS questions*

Looking at the forces in turn, IT/IS can be valuable in building-in switching costs for customers. For example, American Hospital Supplies Company (AHS) established direct IT/IS links with their customers by installing order terminals in hospitals, allowing purchase orders to be placed by less-skilled, low-paid clerks, cutting order costs and providing more flexibility to customers. This resulted in AHS's revenue growing whilst the rest of the industry was in decline, and a return on sales that was four times the industrial average. A further use of IT/IS is the ability to collect and analyse a large amount of information relating to buyers, and establish which buyers are most attractive and which cost the business money. This helps to determine the service level for each buyer.

A similar electronically linked system can be installed between the firm and suppliers, and this enhanced information flow is incredibly useful for supporting systems such as just-in-time (JIT). An example of this is being implemented by Ford Europe: automated bills of materials and vendor systems make it easier to evaluate sources of materials and make or buy decisions. A useful technology in this area is EDI.

Barriers to entry to a market may be erected through the successful application of IT/IS to achieve economies of scale in production. Any new firm wishing to compete must be at least as productive. In addition, computer-aided design and computer-aided manufacturing (CAD/CAM) systems have a major impact on the lead-time for product development, customization and delivery; new entrants wishing to compete in markets where these technologies are prevalent must first invest in the technology.

The threat of substitute products can also be influenced by IT/IS. Flexible manufacturing systems have made it quicker, easier and cheaper to incorporate enhanced features into products, and CAD/CAM permits customization of a product in a way not possible before; providing these technologies are employed by a firm and not its competitors, a distinctive competitive advantage results.

The above implementations regarding new entrants and substitutes may also be used to compete against existing competitors. In summary, the following questions can be used to analyse IT/IS opportunities against suppliers, customers and competitors (including potential entrants and substitutes) (McLaughlin, 1994):

1. *Suppliers* – can we use IT/IS to:
   - gain leverage over our suppliers;
   - reduce buying costs;
   - reduce suppliers' costs;
   - identify alternative sources of supply;
   - improve the quality of products/services purchased?
2. *Customers* – can we use IT/IS to:
   - reduce our customers' costs/increase their revenue;
   - increase our customers' switching costs (to alternative suppliers);
   - increase our customers' knowledge of our products/services;
   - improve support/service to customers and/or reduce cost of satisfying customers;
   - discover more about our customers (and prospects) and their needs?
3. *Competitors* – can we use IT/IS to:
   - raise the entry cost of potential competitors;
   - differentiate or create products/services;
   - make a pre-emptive strike, which prevents competitors' action;
   - reduce our costs;
   - form joint ventures to enter other markets;
   - identify/establish a new market niche;
   - alter the channels of distribution?

7.5 The applications portfolio

7.5.1 General concept

Given the amount of time that information systems have been in common use, the average organization has a variety of different applications supporting different company functions. This list of different applications is known as the *IS application portfolio*. Different applications within the portfolio provide a differing amount of present and potential contribution to the business. Often, applications will differ in age, currency of technology and relevance. Some will be of great strategic importance, whilst others will play a cost-effective and useful but distinctly supporting role. This section looks at ways of organizing the product portfolio and examining the different components in order to assess the way in which they should be managed to best contribute to business strategy.

The *applications portfolio box* is similar to the product portfolio approach (McLaughlin, 1994). It treats applications as products and classifies applications according to the contribution they make, or may make, to the business within the actual and expected competitive environment of the organization. There are four types of classifications defined: support, factory, strategic and high potential, or turnaround. These are illustrated below in Figure 7.6.

Definitions of each of the sectors are shown below:

Support: these are applications that support management effectiveness but are not critical to success. They provide mainly economic benefits. Typical applications in this sector would include payroll, general accounting and employee records.

STRATEGIC	TURNAROUND
Applications which are critical to sustaining future business strategy	Applications which may be important to achieving future success.
Applications on which the organisation currently depends for its success.	Applications which are valuable but not critical to success.
FACTORY	SUPPORT

Figure 7.6 *The application portfolio box*

Factory: factory, or operational, applications are those that are critical to the successful operation of the organization, and without which the organization would suffer serious disadvantage. They support core business activity, and generally include applications such as inventory management, order processing and production control.

Strategic: applications in this sector are those that are critical to the future success of the organization.

Turnaround: these are applications that may be of future strategic importance but are not yet fully developed. They have high potential, but this potential is not yet realized. Expert systems would probably fall into this category.

It is important to realize that the position of certain applications within the portfolio will vary between companies for the same application, depending on the company's business strategy and, to some extent, the industry in which the company is operating. Applications can also drift between sectors. Considering MRP systems, when they were first implemented they would have fallen into the turnaround or strategic sectors, but as they became a more integral part of operations the move around to the factory sector occurred. Generally, there are three driving forces that cause this shift:

- the match between the potential of IT/IS and the firm's operations and strategy;
- the strategic choices that senior management make about IT, e.g., whether to exploit IT/IS to improve productivity, to move into new businesses, etc.;
- changes unfolding in the firm's competitive environment.

Through classifying applications in the above way, it is possible to evaluate the contribution to the business of existing systems and may reveal areas where IT/IS is not being used effectively.

7.5.2 Managing the applications portfolio

Applications in different sectors of the portfolio require management in different ways, despite the fact that in many organizations all applications are managed in a similar fashion. Figure 7.7

shows the different patterns of resource use and benefits generated for applications in different sectors.

Management strategies for each sector can be defined as follows:

7.5.2.1 Strategic

As the objective of applications in this sector is to deliver and sustain business advantage, the commitment and involvement of senior management is required. There is often a need for large investment and, as a result, central co-ordination of requirements is very important. Evaluation of this investment should be based on business benefits rather than simple return-on-investment criteria.

7.5.2.2 Turnaround

Investment in this area is high risk, as there is no certainty that they will deliver the competitive advantage hoped for. There will need to be expenditure on research and development that may not provide positive results, and this should take place in an environment that supports experimentation. High levels of involvement from senior managers are generally not required, but funds and resources need to be committed and the concept must be approved.

7.5.2.3 Factory

It is essential that applications in this area are both reliable and robust, as the company depends on them for daily support of its

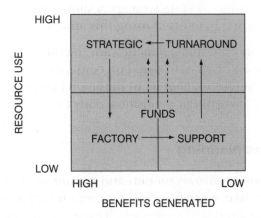

Figure 7.7 *Application management*

core business. Risk reduction is of prime importance, and as a result only proven technologies should be used. The system should have excess capacity to allow all user requests to be met, and centralized control of this resource is demanded.

7.5.2.4 Support

Senior management involvement is not generally necessary, however, tight budgetary controls need to be enforced, and justification for investments in this sector should be financial.

7.6 Critical success factors

An additional way of aligning IT/IS to business strategy is through examining critical success factors (CSF). These are areas in which successful performance will lead to achievement of key objectives, or more simply the few key areas in which things must go right.

Before CSFs can be determined, an organization must decide on its business objectives, a core aspect of determining corporate strategy. Following this, the CSFs required to meet this strategy can be decided. For example, for the business objective of becoming a low-cost producer, critical success factors will include less non-profitable products, less special orders and less non-productive staff. The process for determining CSFs is shown in Figure 7.8.

Although not all CSFs will require IT/IS solutions, this process of analysis may highlight areas where IT/IS will be useful, either as a solution or to monitor performance of the CSF. The process for linking this method to IT/IS strategy is shown in Figure 7.9.

In summary, the benefits of using this analysis method include:

- focusing on what management want the business to achieve;
- providing an IT/IS plan based on consensus;
- involving senior management in the process;
- linking IT investment to business objectives.

7.7 Strategic planning

Having examined various models and techniques that may be used to align IT/IS to corporate strategy, an examination of the practice of strategic planning is necessary. There are six factors that are critical to this:

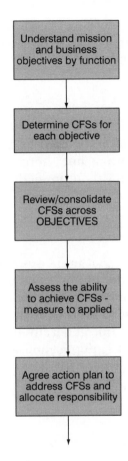

Figure 7.8 *The critical success factor process*

- clear objectives;
- management endorsement;
- partnership approach;
- effective mechanisms;
- honest evaluation and communication;
- learn from experience.

This leads to the six generic strategies that are employed.

7.7.1 Centrally planned

Companies employing this have their IT/IS strategy totally integrated, inter-departmental with corporate strategy. This central control enables an understanding of competitive opportunities and requirements, allows resources to be optimally deployed and large investments to be undertaken.

7.7.2 Leading edge

This entails the maintenance of state-of-the-art technology within the firm. IT/IS can therefore be used to create an advantage over less-developed competitors, but there may be the problem of systems being developed without immediate application. This can be very expensive.

7.7.3 Free market

Working on the principle that users can best determine their own needs, development of the applications portfolio is not centrally controlled. This results in the IS function competing with outside vendors as users attempt to fulfil their own needs in the best way. Senior management are not generally involved with implementation and the end result is duplication and differentials in IT/IS

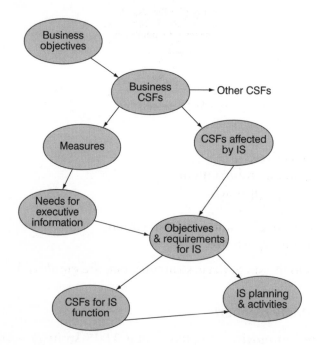

Figure 7.9 *CSFs and information systems*

development rates across the company, along with poor integration of systems.

7.7.4 Monopoly

IT/IS is controlled by data processing, which attempts to meet demand at a reasonable cost. User satisfaction is the measure of success, and there is slow innovation. This also requires excess capacity in order to cope with peaks in demand.

7.7.5 Scarce resource

Use of information systems is heavily constrained by financial considerations, with budgets being set before demand is known. Projects need to be justified by return on investment, not business benefits, and there is a heavy focus on costs rather then value added.

7.7.6 Necessary evil

Only essential use of IT/IS is allowed, either very well cost-justified or statutory. IS management is very defensive and any strategic opportunities for the use of IT/IS are neglected.

7.8 Conclusion

It has been shown how in order to receive strategic benefits from IT/IS it is necessary first for a company to realize the stage it is at with its capability in this area. This can be done through determining at which stage on the IS/IT evolution curve the company is. The areas where application of IT/IS can have the most impact for a company can then be determined, through examining the company internally (using value chain analysis) and through looking at the external competitive environment (using the five forces model).

7.9 References

McLaughlin R (1994). *Information Technology in the Manufacturing Sector*, Elsevier Science, Brussels.

Peppard J (1993). *IT Strategy for Business*, Pitman Publishing, London.

7.10 Further reading

Marks P (1990). Managing computer-aided engineering technology, *NCGA'90 Conference Proceedings*, 2, pp. 109–129.

Woodgate, H.S (1988). *Implementing IT Strategy in a Manufacturing Company*, Elsevier Science, Brussels.

Appendix 1: Basic SQL Commands

CREATE TABLE statement

Creates a new table.

CREATE TABLE table (field1 type [(size)] [**NOT NULL**] [index1] [, field2 type [(size)] [**NOT NULL**] [index2] [, ...]] [, **CONSTRAINT** multifieldindex [, ...]])

Part	Description
table	The name of the table to be created.
field1, field2	The name of field or fields to be created in the new table. You must create at least one field.
type	The data type of field in the new table (see data type section at the end of this document).
size	The field size in characters (text and binary fields only).
index1, index2	A **CONSTRAINT** clause defining a single-field index.
multifieldindex	A **CONSTRAINT** clause defining a multiple-field index.

INSERT INTO statement

Adds a record or multiple records to a table. This is referred to as an *append query*.

Multiple-record append query:

      INSERT INTO target [**IN** externaldatabase] [(field1[, field2[, ...]])]

      SELECT [source.]field1[, field2[, ...]]

      FROM tableexpression

Single-record append query:

      INSERT INTO target [(field1[, field2[, ...]])]

      VALUES (value1[, value2[, ...]])

Part	Description
target	The name of the table or query to append records to.
externaldatabase	The path to an external database. For a description of the path, see the **IN** clause.
source	The name of the table or query to copy records from.
field1, field2	Names of the fields to append data to, if following a target argument, or the names of fields to obtain data from, if following a source argument.
tableexpression	The name of the table or tables from which records are inserted. This argument can be a single table name or a compound resulting from an **INNER JOIN**, **LEFT JOIN** or **RIGHT JOIN** operation or a saved query.
value1, value2	The values to insert into the specific fields of the new record. Each value is inserted into the field that corresponds to the value's position in the list: value1 is inserted into field1 of the new record, value2 into field2, and so on. You must separate values with a comma, and enclose text fields in quotation marks (' '). If you append records to a table with an autonumber field and you want to renumber the appended records, don't include the autonumber field in your query. Do include the autonumber field in the query if you want to retain the original values from the field.

SELECT statement

Instructs the SQL to return information from the database as a set of records.

> SELECT [predicate] { * | table.* | [table.]field1 [**AS** alias1]
> [, [table.]field2 [**AS** alias2] [, ...]]}
> **FROM** tableexpression [, ...] [**IN** externaldatabase]
> [**WHERE**...]
> [**GROUP BY**...]
> [**HAVING**...]
> [**ORDER BY**...]
> [**WITH OWNERACCESS OPTION**]

Part	Description
predicate	One of the following predicates: **ALL**, **DISTINCT**, **DISTINCTROW**, or **TOP**. You use the predicate to restrict the number of records returned. If none is specified, the default is **ALL**.
*	Specifies that all fields from the specified table or tables are selected.
table	The name of the table containing the fields from which records are selected.
field1, field2	The names of the fields containing the data you want to retrieve. If you include more than one field, they are retrieved in the order listed.
alias1, alias2	The names to use as column headers instead of the original column names in table.
tableexpression	The name of the table or tables containing the data you want to retrieve.
externaldatabase	The name of the database containing the tables in tableexpression if they are not in the current database.

WHERE clause

> **SELECT** fieldlist
> **FROM** tableexpression
> **WHERE** criteria

Part	Description
fieldlist	The name of the field or fields to be retrieved along with any field-name aliases, selection predicates (**ALL, DISTINCT, DISTINCTROW,** or **TOP**), or other **SELECT** statement options.
tableexpression	The name of the table or tables from which data are retrieved.
criteria	An expression that records must satisfy to be included in the query results. It can contain up to 40 expressions linked by logical operators, such as **AND** and **OR**.

GROUP BY clause

Combines records with identical values in the specified field list into a single record. A summary value is created for each record if you include an SQL aggregate function, such as **SUM** or **COUNT**, in the **SELECT** statement.

> **SELECT** fieldlist
> **FROM** table
> **WHERE** criteria
> [**GROUP BY** groupfieldlist]

Part	Description
fieldlist	The name of the field or fields to be retrieved along with any field-name aliases, SQL aggregate functions, selection predicates (**ALL, DISTINCT, DISTINCTROW,** or **TOP**), or other **SELECT** statement options.
table	The name of the table from which records are retrieved.
criteria	Selection criteria. If the statement includes a **WHERE** clause, the SQL groups values after applying the **WHERE** conditions to the records.
groupfieldlist	The names of up to 10 fields used to group records. The order of the field names in groupfieldlist determines the grouping levels from the highest to the lowest level of grouping.

HAVING clause

Specifies which grouped records are displayed in a **SELECT** statement with a **GROUP BY** clause. After **GROUP BY** combines records, **HAVING** displays any records grouped by the **GROUP BY** clause that satisfy the conditions of the **HAVING** clause.

> **SELECT** fieldlist
> **FROM** table
> **WHERE** selectcriteria
> **GROUP BY** groupfieldlist
> [**HAVING** groupcriteria]

Part	Description
fieldlist	The name of the field or fields to be retrieved along with any field-name aliases, SQL aggregate functions, selection predicates (**ALL**, **DISTINCT**, **DISTINCTROW**, or **TOP**), or other **SELECT** statement options.
table	The name of the table from which records are retrieved. For more information, see the **FROM** clause.
selectcriteria	Selection criteria. If the statement includes a **WHERE** clause, the SQL groups values after applying the **WHERE** conditions to the records.
groupfieldlist	The names of up to 10 fields used to group records. The order of the field names in group-fieldlist determines the grouping levels from the highest to the lowest level of grouping.
groupcriteria	An expression that determines which grouped records to display. A **HAVING** clause can contain up to 40 expressions linked by logical operators, such as **AND** and **OR**.

ORDER BY clause

Sorts a query's resulting records in a specified field or fields in ascending or descending order.

> **SELECT** fieldlist
> **FROM** table

WHERE selectcriteria
[**ORDER BY** field1 [**ASC** | **DESC**][, field2 [**ASC** | **DESC**]][,
...]]

Part	Description
fieldlist	The name of the field or fields to be retrieved along with any field-name aliases, SQL aggregate functions, selection predicates (**ALL**, **DISTINCT**, **DISTINCTROW**, or **TOP**), or other **SELECT** statement options.
table	The name of the table from which records are retrieved.
selectcriteria	Selection criteria. If the statement includes a **WHERE** clause, the SQL orders values after applying the **WHERE** conditions to the records.
field1, field2	The names of the fields on which to sort records.

Remarks:
The default sort order is ascending (A to Z, 0 to 9). To sort in descending order (Z to A, 9 to 0), add the **DESC** reserved word to the end of each field you want to sort in descending order.

DELETE statement

Creates a delete query that removes records from one or more of the tables listed in the **FROM** clause that satisfy the **WHERE** clause. To drop an entire table from the database, you can use a **DROP** statement.

DELETE [table.*]
FROM table
WHERE criteria

Part	Description
table	The optional name of the table from which records are deleted.
table	The name of the table from which records are deleted.
criteria	An expression that determines which records to delete.

Important notes:
- after you remove records using a delete query, you cannot undo the operation. If you want to know which records were deleted,

first examine the results of a select query that uses the same criteria, and then run the delete query;

■ maintain backup copies of your data at all times. If you delete the wrong records, you can retrieve them from your backup copies.

DROP statement

Deletes an existing table from a database or deletes an existing index from a table.

DROP {**TABLE** table | **INDEX** index **ON** table}

Part	Description
table	The name of the table to be deleted or the table from which an index is to be deleted.
index	The name of the index to be deleted from table.

UPDATE statement

Creates an update query that changes values in fields in a specified table based on specified criteria.

UPDATE table
SET newvalue
WHERE criteria

Part	Description
table	The name of the table containing the data you want to modify.
newvalue	An expression that determines the value to be inserted into a particular field in the updated records.
criteria	An expression that determines which records will be updated. Only records that satisfy the expression are updated.

Sub-queries and multi-table queries (joins)

A sub-query is a **SELECT** statement nested inside a **SELECT**, **SELECT...INTO**, **INSERT...INTO**, **DELETE**, or **UPDATE** statement or inside another sub-query. You can use three forms of syntax to create a sub-query:

comparison [**ANY** | **ALL** | **SOME**] (sqlstatement)
expression [**NOT**] **IN** (sqlstatement)
[**NOT**] **EXISTS** (sqlstatement)

A sub-query has these parts:

Part	Description
comparison	An expression and a comparison operator that compares the expression with the results of the sub-query.
expression	An expression for which the result set of the sub-query is searched.
sqlstatement	A **SELECT** statement, following the same format and rules as any other **SELECT** statement. It must be enclosed in parentheses.

Many useful queries request data from more than one table in a database. SQL allows the user to retrieve data that answers these requests through multiple-table queries that joins data from two or more tables. In relational terms, a join is a combination of three operations: a Cartesian product followed by a selection (row selection), followed by a projection (column selection). After a Cartesian product is formed, we must eliminate all but those rows that join or combine data about the same entity. The selection process will then filter (using the **WHERE** clause) these rows containing data about two or more entities. The last step in carrying out a join is a projection operation, which picks up the required column. Columns in the final result of a join can come from any or all tables named in **FROM** clause.

It is important to realize that queries involving joins can make use of all valid features available for **SELECT** statement, as explained above.

Calculations and comparisons

An SQL can include calculated columns whose values are calculated from the stored data values. The expressions can involve additions (+), subtraction (–), multiplication (*) and division (/). More complex expressions can be built by using parentheses in the usual programming manner. If parentheses are not used, a usual default rule applies: multiplication and division are performed before addition and subtraction.

The **SUM**, **AVG** (**AVERAGE**), **MIN** (minimum) and **MAX** (maximum) commands return corresponding single numeric values of all the non-null data values in the column, and work only with numeric columns. The **WHERE** clause is used to specify comparison conditions.

1. The comparison operators are =, < >, <, < =, >, > =.
2. Range test *between* tests whether the value falls within the specified range of values, while *not between* tests whether the value falls outside specified range of values.
3. Membership test in checks whether the value of an expression falls within the specified values.
4. Pattern matching test *like* checks if the value of a column containing string data matches the specified pattern. The asterisk (*****) wildcard matches any sequence of zero or more characters, while question mark (**?**) matches any single character. *Not like* form of pattern matching can also be used.
5. Multiple search conditions can be specified by using logical **AND** (all conditions must be met), **OR** (any condition can be met) and **NOT** (no condition is to be met).
6. Multiple result tables from separate **SELECT** statements can be combined into a single results table using the **UNION** operation, which combines the results in a manner similar to the operation of **UNION** in set theory.

Equivalent ANSI SQL data types

The following table lists ANSI SQL data types and the equivalent Microsoft Jet database engine SQL data types and their valid synonyms.

ANSI SQL data type	Microsoft Jet SQL data type	Synonym
BIT, BIT VARYING	BINARY (See Notes)	VARBINARY
Not supported	BIT (See Notes)	BOOLEAN
Not supported	BYTE	INTEGER1
Not supported	COUNTER	AUTOINCREMENT
Not supported	CURRENCY	MONEY
DATE, TIME	DATETIME	DATE, TIME
Not supported	GUID	
DECIMAL	Not supported	
REAL	SINGLE	REAL
DOUBLE PRECISION	DOUBLE	FLOAT
SMALLINT	SHORT	INTEGER2, SMALLINT
INTEGER	LONG	INT, INTEGER, INTEGER4
INTERVAL	Not supported	
Not supported	LONGBINARY	GENERAL, OLEOBJECT
Not supported	LONGTEXT	LONGCHAR
CHARACTER	TEXT	ALPHANUMERIC, CHAR
Not supported	VALUE (See Notes)	

Notes:
1. The ANSI SQL BIT data type doesn't correspond to the Microsoft Jet SQL BIT data type, but it corresponds to the **BINARY** data type instead. There is no ANSI SQL equivalent for the Microsoft Jet SQL BIT data type.
2. The **VALUE** reserved word doesn't represent a data type defined by the Microsoft Jet database engine.

Appendix 2: Typical Examination Questions

Chapter 1

1. Briefly describe the term *manufacturing*. What are manufacturing resources?
2. Explain how financial evolution of manufacturing and technical evolution of computers and communication systems affect the need for information. Has cheap communication enabled the use of information, or the need for information in manufacturing evolved the computer and communications systems?
3. Specify, in your own words, the potential benefits of manufacturing information and data systems (MIDS).
4. Explain and contrast the terms information technology (IT) and information systems (IS).
5. Specify and explain major control activities at factory level. Map the IT components for each activity.
6. With a help of the CIM pyramid, explain the concept of CIM. What is vertical and what is horizontal integration within CIM?
7. Describe the term *information* in both a mathematical and physical sense. Which one would suit better in defining manufacturing information? Why?
8. Describe and contrast the terms *human information* and *physical information*. How does communication affect both of these types of information?
9. What is the difference between structural and kinetic information?

10. What is infometrics? Specify the units used to measure information.
11. What are the uses of information? Specify the concomitant value of information.
12. Using the information intensity matrix, place the following products: a car, a bank account, a theatre performance, a TV broadcast. Briefly explain the reason for your choice.
13. Briefly describe five critical characteristics of manufacturing information.

Chapter 2

14. What is a database?
15. Briefly describe the terms physical entity, attribute and attribute domain.
16. A computer file system can be used as a database. What are problems related to such a database?
17. What is a database management system? Specify the functions that a standard database management system needs to perform.
18. Describe how a hierarchical database system works. Give an example. What are virtual records?
19. Describe how a network database works. What is the major advantage over a hierarchical database?
20. List and describe some of the problems of the network database.
21. Describe how a relational database works. What is a structured query language?
22. What is the difference between a logical and physical view of data?
23. What are object-oriented databases? How do they differ from traditional databases?
24. What are hypermedia databases? List the advantages of hyper-media databases over object-oriented databases.
25. Describe and contrast the terms *navigational database* and *relational database* in terms of their physical structure.
26. Describe and contrast the terms conceptual model, logical model and physical model in the database sense.
27. What are entities, relationships and attributes in the relational sense?
28. Define and describe key and foreign attributes in the relational sense.

29. Define the three rules for the design of a relational database. Give an example for each of them.
30. Design a relational database that will represent student details including their names, IDs, modules they take and marks, departments they belong to and details of departmental contact person (name and telephone number).
31. Define and describe reflexive relationships. Give an example.
32. A bill of materials specifies all the components and sub-assemblies required for a product. Using a reflexive relationship(s), design a relational database that will represent a bill of materials.
33. A company wants to create a database of all the parts used in the manufacturing process. The database is to provide information on a proper identification of the parts, price and supplier. Assuming that one supplier generally supplies more than one part, but also that different supplier can supply the same part:
    a) identify the entities, their attributes and the relationship(s) between the entities. Draw up an entity-relationships (E-R) diagram;
    b) specify the primary key for each entity;
    c) design a relational database. Show each table with filled-in examples, and explain the relationships between them. Show which column(s) would be used to design a purchase order.
34. A database is required to support the following requirements at a college:
    a) for a department, store its number and name;
    b) for a lecturer, store his or her name and the number of the department to which he or she is assigned;
    c) for a module, store the module code and the module description;
    d) for a student, store his or her ID, name and the department to which he or she belongs to;
    e) for each module the student has taken, store the module code, module description and the grade received;
    f) for the personal tutor, store his or her name. Assume that one tutor can have more tutees, but each student can have only one tutor.
35. Describe the purpose of database normalization. What is data redundancy and what are the anomalies that result from data redundancy?

36. Describe the following data anomalies: insertion, deletion and modification. How are all of them related to data redundancy?

37. Define and describe functional dependences in relational sense. Given the entity **Student** of the following form:
    Student (Student_ID, Student_Name, Course, Department, Department_Contact)
    where the attribute names are self-explanatory, specify and show all functional dependencies.

38. Define and describe transitive dependencies. Give an example of the entity containing a transitive dependency(ies).

39. Define and describe the first normal form (1NF). Test the entity **Part** of the following form:
    Part (Part_ID, Part_Name, Supplier, Warehouse)
    for the first normal form. The attribute names are self-explanatory.

40. Define and describe the second normal form (2NF). Give an example of the entity that is in the first normal form but not in the second normal form. Show how it can be converted into the second normal form.

41. Describe and define the third normal form.

42. Define and describe the Boyce-Codd normal form.

43. Define and describe a centralized manufacturing database.

44. Define and describe a decentralized database. Contrast its advantages and disadvantages with those of a centralized database.

45. Describe a distributed manufacturing database. Compare it with an ordinary decentralized database.

Chapter 3

46. Define a material requirements planning system (MRP). What are the inputs and outputs of an MRP system?

47. Define and describe a master production schedule. What are firm orders and what are forecasts?

48. Define and describe a bill of materials. What is the difference between a single-level bill of materials and an indented bill of materials?

49. What are inventory records?

50. Define and describe the item master data and the transaction data. Why is the part lead-time important?

51. Specify the advantages of a closed-loop MRP.
52. Specify the main benefits of a successfully implemented MRP II system. What is an ABCD checklist?
53. Give the main steps in implementing a MRP II system.
54. Explain the importance of data integrity for an MRP II system to run successfully.

Chapter 4

55. Define and describe the term *shop-floor data collection*.
56. Why is shop-floor data collection important for a company? Give some examples of data required in a company.
57. What is computerized shop-floor data collection?
58. Describe bar code technology used for shop-floor data collection. How does a bar code reader operate?
59. Specify the advantages and disadvantages of bar codes for shop-floor data collection.
60. Describe electronic labels.
61. Describe and contrast two technologies used for shop-floor data collection: optical character recognition and magnetic strips. What is a voice recognition system?

Chapter 5

62. What is a telecommunications system? What are the principal functions of all telecommunications systems?
63. Name and briefly describe each of the components of a telecommunications system.
64. Distinguish between an analogue and digital signal.
65. Name the different types of telecommunications transmission media and compare them in terms of speed and cost.
66. What is the relationship between bandwidth and the transmission capacity of a telecommunications channel?
67. What is the difference between synchronous and asynchronous transmission? What is the difference between half-duplex, duplex and simplex transmission?
68. Name and briefly describe the three principal network topologies.
69. Briefly describe and contrast the terms PBX and LAN.
70. Define and describe WAN.

71. Define the following: modem, baud rate, gateway and VAN.
72. Name and briefly describe the telecommunications applications that can provide strategic benefits in general and, in particular, competitive advantage to businesses.
73. What are the principal factors to consider when developing a telecommunications plan?

Chapter 6

74. What is the Internet? Explain how independent networks can be connected to enable the Internet to function.
75. Describe and contrast packet-switching technology and the Internet communication protocol.
76. Name and describe the Internet services available today.
77. What is the World Wide Web? Why is the WWW so useful for individuals and businesses?
78. Name and describe major technologies that enable the Internet to operate.
79. How can the Internet facilitate electronic commerce?
80. Describe a typical model for electronic commerce.
81. Describe and contrast the terms *intranet* and *extranet*.
82. Define and describe the five basic rules of electronic commerce.

Chapter 7

83. What is strategic IS?
84. Briefly describe stages of IT/IS evolution since 1950.
85. Briefly describe the basic value chain model. Specify primary and support activities.
86. Explain how a value chain can be applied to an IT/IS strategy. Give some examples.
87. Briefly describe five competitive forces that any industry is exposed to. What is the switching cost?
88. Explain how IT/IS can affect five competitive forces. Give some examples.
89. What is an application portfolio? Specify strategic, factory, turnaround and support applications.
90. What are critical success factors?
91. Explain major factors in strategic planning.

Background and Rationale of the Series

This new series has been produced to meet the new and changing needs of students and staff in the Higher Education sector caused by firstly, the introduction of 15 week semester modules and, secondly, the need for students to pay fees.

With the introduction of semesters, the 'focus' has shifted to module examinations rather than end of year examinations. Typically, within each semester a student takes six modules. Each module is self-contained and is examined/assessed such that on completion a student is awarded 10 credits. This results in 60 credits per semester, 120 credits per year (or level to use the new parlance) and 360 credits per honours degree. Each module is timetabled for three hours per week. Each semester module consists of 12 teaching weeks, one revision week and two examination weeks. Thus, students concentrate on the 12 weeks and adopt a compartmentalized approach to studying.

Students are now registered on modules and have to pay for their degree per module. Most now work to make ends meet and many end up with a degree and debts. They are 'poor' and unwilling to pay £50 for a module textbook when only a third or half of it is relevant.

These two things mean that the average student is no longer willing or able to buy traditional academic text books which are often written more for the ego of the writer than the needs of students. This series of books addresses these issues. Each book in the series is short, affordable and directly related to a 12 week teaching module. So modular material will be presented in an accessible and relevant manner. Typical examination questions will also be included, which will assist staff and students.

However, there is another objective to this book series. Because the material presented in each book represents the state-of-the-art practice, it will also be of interest to professional engineers in industry and specialist practitioners. So the books can be used by engineers as a first source reference that can lead onto more detailed publications.

Therefore, each book is not only the equivalent of a set of lecture notes but is also a resource that can sit on a shelf to be referred to in the distant future.

Index